PLANTS

OF

YELLOWSTONE

AND

GRAND TETON NATIONAL PARKS

including the Greater Yellowstone Ecosystem

Richard J. Shaw
Professor Emeritus and Emeritus Director
of Intermountain Herbarium, Department of Biology,
Utah State University, Logan, Utah

Wheelwright Publishing
373E Forest Glen Lane
Camano Island, WA 98282
Tel: (360) 387-0960

Calliope Hummingbird visits Columbia Monkshood Photo by Willard Dilley

Wheelwright Publishing
373E Forest Glen Lane
Camano Island, WA 98282
ISBN 0-9702067-0-4 (previously ISBN 0-937512-02-8)

Picking wildflowers or collecting specimens of animals, trees or minerals in all National Parks and Monuments is prohibited without special permission from the park superintendent. Study the plants where they grow, take home photographs of them, but leave them for the enjoyment of those who will follow.

Front Cover — Western St. Johnwort (see page 100)
Photo by Danny On

TABLE OF CONTENTS

INCHES 1 2 3 4 5 6

METRIC 1 2 3 4 5 6 7 8 9 10 11 12 13 14 15

METRIC SYSTEM TABLE

1 mm. = approx. 1/25 of an inch
10 mm. = 1 cm. (approx. 2/5 of an inch)
10 cm. = 1 dm. (approx. 4 inches)
10 dm. = 1 m. (approx. 40 inches)

INTRODUCTION

The purpose of this plant book is to provide park visitors who lack botanical training a source of information about the fascinating vegetation that blankets the surface of two of North America's most beautiful National Parks. It is the result of many years of study of the plants of the region and consideration of what the visitor wants to know about "that plant." It is hoped that the colored photographs, non-technical descriptions, and convenient size will make the book useful to the backpacker, mountain climber and highway traveler.

Two recent checklists of both parks reveal that over 1,000 species of vascular plants inhabit an incredible number of diverse habitats. Of these many plants, 213 species have been chosen for illustration and description. It will be apparent that many admirable species have been neglected, but an attempt has been made to include all of the tree species, the majority of the shrubs (except for the numerous willow species) and a representative view of the 89 families within the two park boundaries. The combined size of Yellowstone and Grand Teton National Parks exceeds 3,900 square miles and includes an awe-inspiring landscape of thermal areas, glaciers nestled in cirques, craggy peaks above 12,000 feet, mountain lakes of all sizes and river drainages flowing to two oceans. This area, bigger than the state of Delaware, has a cold climate with deep drifting snow, a low mean annual temperature of about 33° F., and a growing season of approximately 60 frost free days.

Arrangements of the Plants

The Table of Contents reveals that aside from the trees and some shrubs the plants in this book are grouped according to color. In a number of cases, particularly in the pinkish, purplish and bluish sections, it was difficult to assign certain species which have color variations. It is advisable to check more than one section if color deviation is suspected. Scientific and common names are included in one index.

Common Names

Long ago botanists reached international agreement that there should be only one valid scientific name for each plant species, but no such action has been taken concerning common names. Indeed, it is unlikely that such action will be taken. Consequently, a multiplicity of common names may occur for many species, often varying from state to state and country to country. For example, *Erythronium grandiflorum* in Utah is most often called Dogtooth Violet, and yet in Colorado it is called Glacier Lily. To solve the problem in this volume, the author has often included several names and tried to follow the names applied by recent identification manuals (see selected refernces).

Brief History of Plants of Yellowstone and Grand Teton National Parks

Field and laboratory studies conducted in both parks in the last two decades tell a story of vegetation spanning millions of years. The vegetational history will never be completely understood, but because of repeated volcanic activity, we can today look at beautifully preserved plant fossils that give us glimpses of climate and plant species of the past. Understanding the past may help to preserve our present ecosystem from foolish mistakes in the future.

Careful study of hundreds of fossils in both parks, including petrified wood, fossilized leaves, cones, seeds and pollen has revealed a succession of forests composed of more than 100 different species of plants. These species include one doubtful cycad, 10 ferns, 3 horse-tails, 10 conifers, (including redwoods) and 76 flowering plants. Many of these better fossils go back to the Eocene Epoch which is estimated as having begun approximately 58 million years ago and continued until 36 million years ago.

At one place in the Lamar Valley of Yellowstone, erosion has exposed petrified stumps of trees of 27 distinct layers of buried forests, one on top of the other. Fossil forests are well known in many countries throughout the world, but in this particular region

Alpine zone at Timberline Lake, Grand Teton

of Yellowstone the alternation of volcanic activity and quiet periods of forest growth were repeated at least 27 times during thousands of years. The petrified wood of the tree stumps is so perfectly preserved that anyone with a hand lens can see growth rings and delicate microscopic details of the wood cells.

Climatic conditions during the Eocene are interpreted as being essentially similar to those now found in southeastern and south-central United States. A warm-temperate to sub-tropical situation existed depending on elevation. Rainfall probably averaged between 50 to 60 inches per year. It is believed that elevation at this time was most likely no more than 3,000 feet above sea level. Today, of course, the average elevation in the parks is about 7,000 feet and rises well above 13,000 feet. It is interesting to compare the forests now living in the parks with those represented in its fossil forests. Evergreen conifers such as pine, spruce and fir dominate today with only a few species of deciduous hardwoods such as cottonwood and aspen. The reverse relationship existed in the Eocene with the hardwoods (maple, magnolia and sycamore) being dominant.

Vegetational patterns of Yellowstone and Grand Teton regions now are indicative of a cool-temperate to subarctic climate. The change from a subtropical climate some 50 million years ago to the present cold climate has been the result of world wide lowering of temperature coupled with an uplifting of the entire Rocky Mountain region nearly 7,000 feet during the last 36 million years. At least three times in the last 250,000 years glaciers from surrounding highlands have swept across both parks' landscapes destroying the vegetation in their paths. A few of the highest mountains were islands in the ice flows and served as a place of refuge for some species of plants.

Petrified Tree Stump at "Fossil Forest," Yellowstone

In 1970 a scientific paper gave some additional insight into rather recent vegetational changes. A series of depressions at the end of the southeast arm of Yellowstone Lake were drilled for pollen and seed analysis. The present vegetation surrounding the site where the core was taken is Lodgepole Pine *(Pinus contorta)* forest. Careful study of the cores records the vegetation sequences from the retreat of Late Wisconsin (Pinedale) ice to the present time. Richard C. Baker, author of the paper, suggests that the vegetation was a subalpine parkland or alpine tundra near the tree line during late glacial time, including spruce, fir and White-bark Pine. The inferred vegetation changed rapidly to a Lodgepole Pine forest shortly before 11,500 years ago. This significant change indicates a distinct warming trend at that time. The cores further suggest that the *Pinus contorta* forest has persisted with minor modification ever since.

Other authorities of plant migration in western North America have suggested that the warming trend mentioned above allowed the invasion of plants adapted to dry conditions from the semi-arid Great Basin to the south and west, possibly by way of the Snake River drainage. The invasion of plants adapted to wetter conditions is more difficult to pinpoint, but there is some evidence that such plants migrated from the Pacific Northwest and the Northern Rocky Mountains. Additional studies like the one mentioned above will surely expand our fascinating view of plant migration, extinction and plant evolution.

Hints on Photographing Wildflowers

The vast majority of wildflowers can be photographed in striking color with today's moderately priced automatic cameras equipped with attachable close-up lenses. Such a lens or series of lenses (which usually retail between $5.00 and $30.00) will allow you to shoot as close as 6 inches from the blossoms. A small tripod will provide the steady support necessary for shutter speeds of 1/50 of a second or more.

You will be happier when you project your slides if you have taken time to compose each picture for maximum detail and appropriate background. Backgrounds may be provided by natural shadows or even blue sky if a low shooting angle is possible. Natural objects in the plant's environment such as a downed log may be helpful and often give a feeling of wilderness. However, there may be times when you want to show accurate flower structure with maximum contrast, and this is the time to use an artificial background. A piece of black cloth or paper held behind the flower may be all that is needed. As insurance, take several pictures of the choice subjects using different angles and exposure times. A portait lens greatly reduces the depth of focus, especially below 18 inches; therefore select plants where the flowers are in the same general plane. If you are not using a single-lens-reflex camera, measure the flower to lens distance to assure sharp focus.

A light meter is advisable for those who wish to conserve film and obtain the finest transparencies. If you do not have one, consult the photo guides provided by your film supplier. You should feel free to ask members of the Naturalist Staff about native plants and their names. With a little practice and patience a photographic effort on wildflowers will be rewarding and will return pleasant memories many times.

ACKNOWLEDGEMENTS

I offer my sincere thanks to R. Alan Mebane, Chief Naturalist of Yellowstone National Park, and Charles McCurdy, former Chief Naturalist of Grand Teton National Park. Without their encouragement and support this volume could not have been completed.

I am particulary grateful to John W. Stockert of the National Park Service for use of ten of his choice photographs and much aid in selecting the most appropriate common names for each plant. Stanley G. Canter, Assistant Chief Naturalist of Yellowstone National Park, and Dr. Don G. Despain, Research Biologist of Yellowstone National Park, also made valuable suggestions.

Dr. Don Despain, Robert Kretzer, Lamar Lane, Margaret Mortensen kindly loaned photographs, and credits for these are given in the text. Clyde M. Lockwood, Chief Naturalist of Glacier National Park, and Executive Secretary of the Glacier Natural History Association, generously allowed the use of 27 transparencies from the Danny On collection (slides are marked DO). The remaining photographs were taken by the author.

Special appreciation goes to my wife, Marion, who spent many hours waiting for photographs to be taken and correcting the manuscript.

KEY TO THE CONIFEROUS TREES ‡

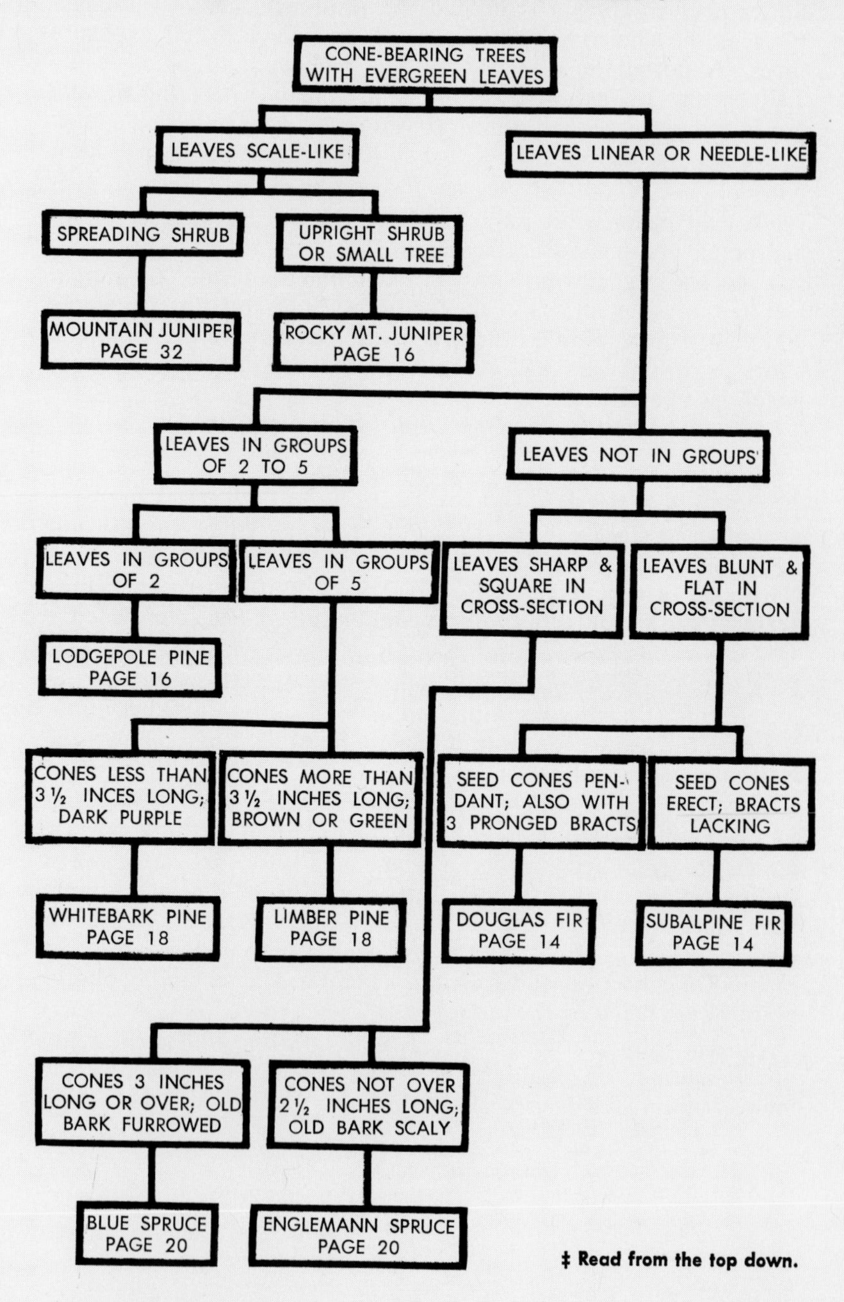

‡ **Read from the top down.**

KEY TO THE DECIDUOUS TREES ‡

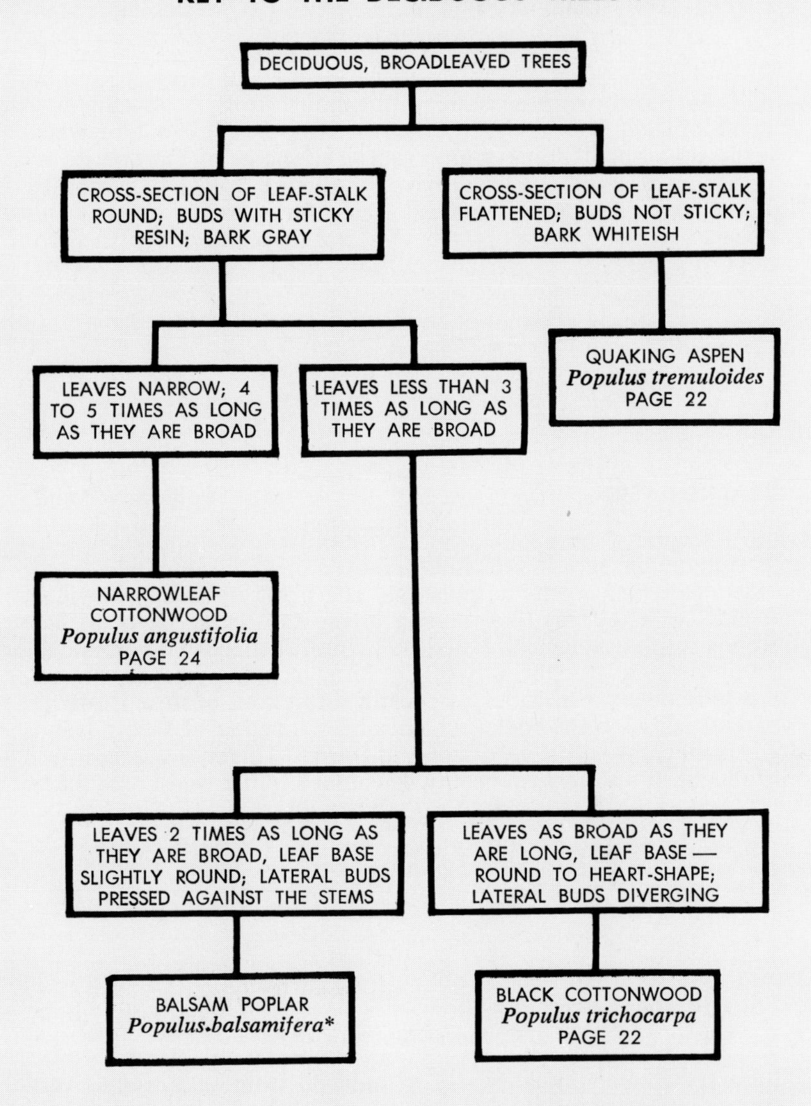

*Some botanists feel that this species does not occur in Yellowstone or Grand Teton, but occurs east and north of our area. Part of the problem is due to the relative ease with which the *Populus* species cross with each other.

‡ **Read from the top down.**

PEPPERWORT; CLOVER-FERN Pepperwort Family
Marsilea vestita

Resembling a 4-leaf clover, this aquatic fern is apparently rare, but its beauty and uniqueness are worth the effort of looking for it. A pond which tends to dry up by August is best for a fern whose leaflets may be floating or emergent. The spores of this species are tightly packed into a small, ovate structure called a sporocarp. These sporocarps may remain dormant for years before they split open to release the spores. About a dozen species of ferns exist in the two parks.

BRACKEN FERN Polypody Family
Pteridium aquilinum

This perennial fern inhabits burned-over sites in forests which provide shade and adequate moisture. The plant starts the growth of fronds in June, and they remain green until early October. The fronds reach a height of 2 to 4 feet and develop from deep underground stems. All portions of the fern, including the dry fronds, are poisonous to livestock, especially cattle and horses. However, several edible plant books recommend the eating of young fronds when they are still shaped like fiddlenecks. This is the largest fern of the parks and its lush beauty adds much to the wilderness scene.

ROCKBRAKE; PARSLEY FERN Polypody Family
Cryptogramma crispa

Fern species inhabit a variety of habitats from very wet to very dry situations, and Rockbrake is one which is adapted to the rather dry cliff crevice and talus of the mountains. The leaves are densely tufted on a short rhizome, and the species can be readily recognized by the short, sterile fronds mixed with the taller fertile fronds. If one looks at the small leafllets of the fertile fronds with a hand lens, the clusters of sporangia will be seen protected by the recurved leaf margin.

Pepperwort 1x

Bracken Fern 1x

Rockbrake ¾x

SUBALPINE FIR Pine Family

Abies lasiocarpa

Perhaps the best way to identify this conifer from a distance is to watch for its long, slender, narrowly-conical crown. When close to this species, one can see that its flat, flexible needles and erect cones distinguish it from all other evergreens. Although Subalpine Fir is a relatively small tree and of little commercial importance, it is the most widely distributed fir in western North America. As the name implies, this species grows from 6,800 feet up to timberline at 10,000 feet. Within the parks it is associated with Engelman Spruce and Lodgepole Pine, especially tolerating the shade of the latter species. The smooth, gray bark of young trees is unique because of the many lens-shaped blisters, often up to 1 inch or more in length, which are filled with a sticky resin. On older trees the bark is little broken except for shallow cracks near the base of the trunk. As the result of heavy snows, the lower branches often become rooted, thus forming a circle of smaller trees. Such a colony is called a snowmat of Subalpine Fir.

DOUGLAS FIR Pine Family

Pseudotsuga menziesii

Douglas Fir is widely distributed in the Western States, reaching its largest size in the coastal region of Washington and Oregon. In the parks of the Rocky Mountain Region, we have a smaller, slower growing variety. This species bears strong resemblance to spruce, fir and hemlock, thus botanists gave it a generic name of *Pseudotsuga* (False Hemlock). The flat, flexible needles are borne singly and grow around the branch giving it a full appearance. The seed cones set the tree apart from the true firs because they hang down and do not disintegrate on the tree as they do in the case of the firs. The best distinguishing feature of this species is found in the female cone. Between the cone scales are prominent three-pronged bracts. As the color plate indicates, these "Neptune's tridents" appear in both early and mature stages of cone development. On the same tree bright red male cones appear in the early spring, but fall off as soon as the pollen is shed. The bark on older individuals is reddish-brown and deeply fissured, often up to 5 inches or more in thickness. Because of the thick bark, some individuals may survive hot forests fires.

Subalpine Fir

Subalpine Fir ½x

Douglas Fir

Douglas Fir ¾x

LODGEPOLE PINE

Pine Family

Pinus contorta

Of the three species of pines found in the two parks, this is the most common, and the only one with 2 needles in a cluster. It is a small tree which seldom gets over 75 feet high. When it grows in dense stands, it is characteristically tall and slender, losing the lower branches as they become shaded. If the individual trees are widely spaced, they become quite bushy and densely branched. The Lodgepole Pine seedlings grow quickly in mineral soil of a fire-cleaned area or in any kind of a disturbed site, particularly road side cuts. The Mountain Pine Bark Beetle is the most serious threat to this species since the larvae of the beetle bore under the bark and eventually girdle the tree. Up to 40 to 50 percent of the trees over 6 inches in diameter have succumbed to the beetle in some areas. Other threats to its existence include the Dwarf Mistletoe and the Comandra Rust. The western Indians used the slender trunks of the small trees as a framework for their teepees or lodges, hence the common name, Lodgepole Pine. From the color plate it is possible to recognize the mature male cones which produce great quantities of pollen in June, the young, green female cone, and the brown, fully-matured female cone containing seeds.

ROCKY MOUNTAIN JUNIPER

Cypress Family

Juniperus scopulorum

Rocky Mountain Juniper is prevalent in the western United States and can be seen easily on dry slopes in the vicinity of Mammoth Hot Springs. It is generally a bushy shrub or small tree, often with several trunks. The leaves are mostly scale-like and opposite; however, some juvenile, needle-like leaves are often persistent until near maturity of the trees. Some individuals produce only pollen cones while others produce only seed cones. The Indians ate the berry-like seed cones in the late summer or fall, and in times of food shortages they even consumed the inner bark.

Lodgepole Pine

Lodgepole Pine ¾ x

Rocky Mountain Juniper

LIMBER PINE Pine Family

Pinus flexilis

This five-needled pine is seldom found in pure stands, but is more often found as a lone individual on the dry, rocky moraines of the valley of Jackson Hole. It is common near Mammoth Hot Springs. The young branches of this tree are very flexible and can be tied in knots without breaking. This peculiar characteristic is of advantage in withstanding the severe winds and is also responsible for both the common and scientific names. In June this species is conspicuous because of its numerous reddish clusters of pollen-bearing cones. The seed-bearing cones produce large seed crops at irregular intervals and these are sought by birds and rodents who serve as important agents of dispersal. Each needle is 1½ to 3 inches long and closely pressed into clusters of five. Such needle clusters often remain on the branches for 5 or 6 years.

WHITEBARK PINE Pine Family

Pinus albicaulis

Since this species is often confused with the Limber Pine, it would be well to compare the color and length of the seed cones of both species. Whitebark Pine seed cones are short (ranging from 2 to 3½ inches) and remain purple to maturity. In contrast, the Limber pine bears seed cones which are longer (ranging from 4 to 8 inches) and remain green until maturity. The bark of Whitebark seedlings is covered with a fine white coating, and the larger trunks sometimes have a whitish cast. Because of this, the scientific name *Pinus albicaulis* has been applied which means literally "the pine with the white stem." This species is important in subalpine zones from 8,000 to 10,000 feet, especially where mountain soils are shallow. Two miles beyond the forks leading into South Cascade Canyon of Grand Teton, grows one of the world's largest Whitebark Pines. This specimen is 18½ feet in circumference and is probably 400 years old. While there are many large individuals in this area, two miles farther at timberline the trees become stunted and dwarfed, attesting to the severity of wind and low temperature at 10,000 feet. This tree is easily seen at Dunraven Pass in Yellowstone.

Limber Pine

Limber Pine ¼x

Whitebark Pine

Whitebark Pine ½x

ENGELMANN SPRUCE Pine Family

Picea engelmannii

Engelmann Spruce inhabits the canyons of the Teton Range above 6,800 feet. In Yellowstone, it grows in shaded, moist ravines in such places as Kepler Cascades, Spring Creek and the South Entrance. In some areas to the south of the parks it reaches elevations of 10,000 to 12,000 feet. The needles and cones are its distinguishing features. Remember the word spruce begins with "s" and the single needles are square in cross-section and sharp to the touch. Check the square in cross-section character by rolling a needle between thumb and fore-finger. The cones have papery scales and resemble the Blue Spruce but are only 1½ inches long, whereas the latter reach 3 inches in length. Perhaps the largest individuals of this species are located near Hidden Falls in Cascade Canyon. Here one will find specimens over 80 feet in height, and at breast level, diameters extending to 36 inches. Such trees may be 400 years or older.

BLUE SPRUCE Pine Family

Picea pungens

While the Blue Spruce seems to be lacking in Yellowstone, it is common along the Snake River as it winds its way through Jackson Hole. The distribution of this widely admired tree is limited to the Central Rocky Mountain Region, and within Grand Teton it will seldom be found above 6,800 feet. The stiff, four-angled needles are sharp to the touch and are up to 1¼ inches in length. The bluish to silvery-green appearance is due to a fine powdery substance on the surface of the needles. Each leaf is borne on a brown, stalk-like projection which is persistent years after the needles fall off. Blue Spruce can be readily distinguished from the Engelmann Spruce in two ways: the seed cones of the Blue Spruce are nearly twice the length of Engelmann Spruce, and the bark of the former is ashy-brown and divided into vertical ridges, while the latter lacks the ridges and is more scaly. The small, winged seeds of spruce are eaten by several kinds of birds, as well as squirrels and chipmunks.

Engelmann Spruce

Engelmann Spruce ½x

Blue Spruce

Blue Spruce ½x

QUAKING ASPEN Willow Family

Populus tremuloides

This is a true poplar in the Willow Family and is called Quaking Aspen because its leaves move in the slightest breeze. This almost incessant trembling of the leaves is due to the slender, flattened, leafstocks. Quaking Aspen is by far the most common broad-leaved tree in the two parks. In fact, it is one of the most widely distributed trees in North America. In the Intermountain Region its reproduction is by means of shoots which come up from horizontal roots. The small, wind-borne seeds are scattered far and wide but seldom find a suitable place for germination. Horizontal ridges and scars mark the thin, white or yellowish-green bark. Bases of old trees are roughened and black. Leaf mining insects are especially common on the leaves of aspen. The larva of the leaf miner lives and feeds between the two epidermal layers of the leaf. During June and July these worm-like, immature stages create intricate patterns on the leaf. In August the larva goes into a resting stage by rolling up in the leaf margin. After a period of dormancy, a small moth emerges from the pupa.

BLACK COTTONWOOD Willow Family

Populus trichocarpa

The Black Cottonwood is the largest broadleaved tree in the two parks; in fact some specimens on Second Creek in Grand Teton reach a diameter of 3 feet. The distinguishing features which set this tree apart are the sticky resin on the buds, the large, broadly ovate leaves which are often up to 5 inches in length, and the deeply furrowed gray bark. The leaves are finely toothed at the margin, dark green on the upper surface and whitish underneath. Flowers of both sexes are borne separately on different trees. By August the more loosely flowered pistillate catkins are 6 to 11 inches long with pointed capsules. The tiny fragile seeds have attached fluffy fibers which carry them long distances. Cottonwoods also reproduce from stumps and root sprouts. The deeply fissured bark of mature trees is 2 to 3 inches thick and provides protection against fire. The narrow-leaf cottonwood *(P. angustifolia)* differs from the Black Cottonwood in having narrow, lance-shaped leaves.

Quaking Aspen

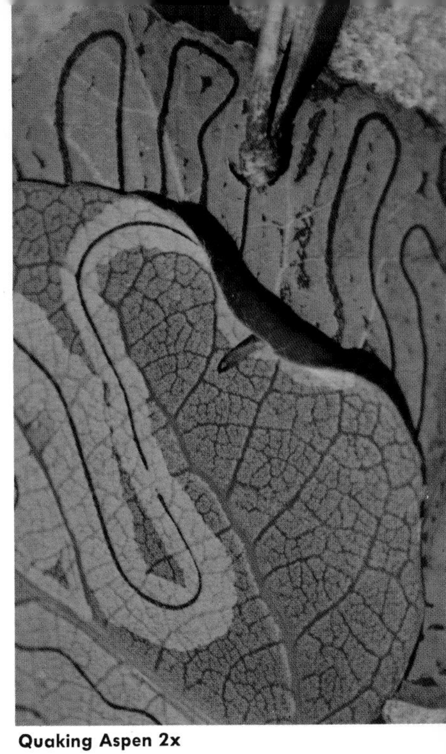
Quaking Aspen 2x

Black Cottonwood

Black Cottonwood ¾x

NARROWLEAF COTTONWOOD Willow Family
Populus angustifolia

Narrowleaf Cottonwood is typically a small, slender tree appearing more like a willow than a cottonwood, mostly because of long, narrow, willow-like leaves. Hybridization with other poplars produces many individuals with intermediate features. The leaves are usually 2 to 3 inches long and ½ to ¾ inches wide. Fine teeth are present on the margins. The hairless fruits are broadly egg-shaped and split into two parts to discharge the seeds. Owing to its intolerance of shade, it is one of the first plants to establish on gravel and sand bars.

BOG OR DWARF BIRCH Birch Family
Betula glandulosa

This shrubby birch is found occasionally along streams in the canyons of the Tetons and south of Old Faithful near the Firehole River Bridge. Like the alders, the birch flowers are small, inconspicuous and borne in slender catkins. After wind pollination has occured, seed-like nutlets develop on the female catkins. Grouse feed upon the catkins and seeds. Moose and beaver are reported to feed upon the young resinous twigs.

THINLEAF; MOUNTAIN ALDER Birch Family
Alnus incana

The alders are tall shrubs or small trees of up to 20 feet in height, and they are found along rivers and streams or in moist meadows. The alternate, egg-shaped leaves have a doubly-toothed leaf margin. The flowers appear early in the spring before the leaf buds open. The pistillate and staminate flowers arise in separate clusters called catkins. The individual blooms are minute and pollination is accomplished by the wind. Because of persistent, woody scales on the female catkin, the entire structure looks like a miniature pine cone. Beavers cut the stems of alders for building their dams and lodges as well as using the bark for food.

Narrowleaf Cottonwood ¾x

Bog Birch ¾x

Thinleaf or Mountain Alder ½x

ARCTIC WILLOW Willow Family

Salix arctica

About 75 species of willows grow in North America and at least 20 species survive in the two parks. Willows are usually thought of as requiring a moist site along riverbanks and streams, but this very small species is successful on well-drained mountain soils around alpine lakes. The inconspicuous flowers appear soon after the snow melts in July. They are arranged along slender, erect stems. The fruits are small pods that split into two parts, releasing minute seeds surrounded by tufts of long, white, silky hairs. The species illustrated has small, leathery leaves to reduce water loss in the harsh alpine environment.

WHIPLASH WILLOW Willow Family

Salix lasiandra

The willows are trees or shrubs with alternate, simple leaves. As a group they are easily recognized by their characteristic twigs and catkins, yet it is often difficult to distinguish between many of the species. The flowers are uni-sexual, the staminate and pistillate ones are both arranged in catkins but on different plants. The Whiplash Willow, illustrated here, is a spreading shrub 3 to 15 feet high. It is abundant at lower altitudes especially along the Snake River. The pistillate catkin in this species is up to 2½ inches in length, and by the time the fruits ripen, the numerous seeds are hairy. Willows are important in controlling erosion along stream banks and in providing food for many species of wildlife. The buds and small, tender portions of the twigs are staple food for several species of grouse. The foliage and twigs furnish browse for moose, elk and deer. The bark is one of the favorite foods of the beaver.

RUSSET BUFFALOBERRY Oleaster Family

Shepherdia canadensis

This thornless shrub is often found in the shade of the Lodgepole Pine forest and along the Snake River. The flowers appear in June and are inconspicuous as the petals are missing. The attractive red-orange berries brighten the branches by mid-August, but do not be misled into eating the fruit as it is bitter and the taste lingers on. The Silver Buffaloberry of northeastern United States has fruit with a pleasant, tart flavor and was used in the early days to garnish buffalo steaks. The upper surface of the leaves is dull green, while the lower surface is silver with rusty patches.

Arctic Willow ¾x

Whiplash Willow 1x

Russet Buffaloberry ¾x

RED RASPBERRY Rose Family

Rubus idaeus

The native Red Raspberry resembles the cultivated Red Raspberries of our gardens. It is commonly found above 7,000 feet on rocky talus slopes. The plant has compound leaves with 3 to 5 leaflets, and the erect stems are prickly. The white blossoms are ½ to ¾ inch across and continue to appear until mid-August. The red berries which ripen from late July to September are sweet and edible, being quite comparable to those of the cultivated raspberries. The fruits are eaten by species of birds and mammals. It is interesting to note that this difficult genus, *Rubus,* has over 200 species in the United States, but the two parks are known to have only two species.

SILVERBERRY Oleaster Family

Elaeagnus commutata

Flowers and fruits of this shrub are not conspicuous, but the silvery pubescence which covers the leaves makes this species unique and easy to recognize. While silverberry is uncommon in Yellowstone, it is frequently seen by those visitors who float the Snake River in Grand Teton. Each flower has a tubular shape, being silvery on the outside and yellowish on the inside. Like the buffaloberry, the flowers lack petals. Note how the remains of the flower hang on to the end of the ½ inch long silvery fruit. The berries are dry and mealy, thus making them unpleasant to the taste.

GROUSE WHORTLEBERRY Heath Family

Vaccinium scoparium

In the shade of the Lodgepole Pine, this low shrub seldom grows more than 12 inches high. The broom-like branches are green and covered with egg-shaped leaves up to ½ inch in length. The nodding flowers are urn-shaped and vary from white to pink. The juicy, red fruits are attractive to many birds and mammals, including man. Many recipes are available to mix the fruit with various kinds of dough to make muffins and pancakes. The Indians also had a variety of ways to use them.

Red Raspberry ⅔ x

Silverberry ¾ x

Grouse Whortleberry ¾ x

DO

BLACK HAWTHORN Rose Family

Crataegus douglasii

Hundreds of names have been proposed for the many species of Hawthorns, and as a group they are perhaps the most difficult to classify in the Rose Family. The generic name, *Crataegus,* is a Greek word meaning strong, in reference to its tough wood. The genus can be readily distinguished even in winter because no other rosaceous tree or shrub has the long, simple thorns (up to one inch in length). The white flowers are borne in clusters at the ends of the branches. By late August the small apple-like fruits have reached maturity and provide food for many birds and small animals. The Indians of some regions gathered the fruits or "haws" and dried them for winter use.

POISON IVY Sumac Family

Rhus radicans

Poison Ivy is rare in the two parks. It has been known to inhabit the Gardiner River, north of Mammoth, for several years, and in 1963 it was found for the first time on the west shore of Jackson Lake. The plant is a sub-erect shrub or woody-stemmed vine which is variable in habit of growth. The best identifying feature is the presence of the shiny, 3-foliate leaves, mostly oval to lanceolate with entire margins. The small, greenish-white flowers give rise to dull, yellowish, waxy fruits, ¼ inch in diameter. Poison Ivy contains a substance chemically similar to lacquer. This resin is non-volatile and cannot produce dermatitis unless a person actually contacts the active compound. The allergin may be carried to the person on particles of carbon in smoke, or by pets, clothing or tools. The resin clings to the skin and may be removed by washing with a strong soap 5-10 minutes after contact. Certain ointments may relieve itching.

Black Hawthorn 1x

Poison Ivy ⅓x

Poison Ivy ½x

ALDER BUCKTHORN Buckthorn Family

Rhamnus alnifolia

This is a low, spreading shrub which is readily distinguishable in summer by its finely toothed leaves, and in winter by its scaly buds. Each leaf has 6 to 8 pair of prominent veins. The small flowers are greenish - yellow, and the petals are generally absent. The berry-like fruit becomes blackish and contains 2 to 3 small nutlets. The closely related tree of the northwest Pacific Coast, *Rhamnus purshiana* (Cascara) was utilized medicinally by several Indian tribes. They prepared a laxative drink by boiling the bark in water. It is found along rivers and streams.

WESTERN WINTERGREEN Heath Family

Gaultheria humifusa

This prostrate evergreen shrub is often overlooked and ignored until the mature fruits ripen in late August. The red berries make an acceptable jam with a spicy odor. The oval, shiny leaves reach up to ¾ inch long, and when they are young and fresh, the mountain hiker will find they are pleasant to chew. Leaves used fresh or dry as a tea have a menthol flavor. The white flowers have a vase-shape and vary between ⅛ to ¼ inch long. Look for the plant in moist soil near bogs and lakes.

COMMON JUNIPER Cypress Family

Juniperus communis

This widely distributed species is to be found across North America, Europe and Asia. Like other junipers, it is easily distinguished by the soft, blue, berry-like fruit. These are not true berries but modified cones whose scales are relished by birds, but the seeds inside are not digested and pass on through the digestive tract. For this reason junipers have a wide distribution. Common Juniper is a spreading, creeping shrub, seldom over 3 feet in height. The sharp-pointed, awl-like needles are in groups of three, ⅜ to ½ inch long. Where the leaf joins the branch, there is a definite constriction. Rocky Mountain Juniper *(J. scopulorum)* differs from the Common Juniper in being a small tree and having scale-like leaves appressed to the twigs.

Alder Buckthorn 1/2x

Western Wintergreen 1x

Common Juniper 1x

WHITESTEM GOOSEBERRY Gooseberry Family

Ribes inerme

Reaching a height of 4 feet, this prickly, slender-stemmed shrub will be found along streams in rather open forest sites. The greenish or purplish flowers give way to the formation of purple-black berries supporting a very persistent calyx at the end opposite the pedicel. Because plants of the genus *Ribes* are alternate hosts for the White Pine Blister Rust, special programs of eradication have reduced the number of these shrubs in many western forests.

ROCKY MOUNTAIN MAPLE Maple Family

Acer glabrum

This species of maple is a slender-stemmed shrub usually less than 12 feet in height. It is found near Mammoth Hot Springs and the lower eastern slopes of the Teton Mts. The 3 to 5 lobed leaves vary from 2 to 3 inches across and are supported by reddish, slender petioles. Greenish-yellow flowers appear with the unfolding leaves in loose drooping clusters. The colorful double-winged fruits mature by mid-August; the seed portion is strongly wrinkled and indented. Many botanists think this maple is suitable for use as an ornamental in small gardens because of its diminutive size and attractive fall coloring.

HIGHBUSH HUCKLEBERRY Heath Family

Vaccinium membranaceum

This bushy shrub varies from 2 to 4 feet high, and it is very popular with park visitors because of its sweet, juicy berries. In southern Yellowstone and Grand Teton it is quite a common inhabitant of the Lodgepole Pine forest and extends up into the canyons of the Tetons as high as 8,500 feet. The globe-shaped flowers are greenish-white or pinkish. The 5 petals are united into a tube and bear 10 stamens on the inner surface. By mid-August the wine-red to purple-black berries begin to appear, and there is considerable competition between birds, bears and man for, perhaps, the most tasty fruit of the parks. There is considerable variation in the size and shape of the fruit. Deer browse freely on the plants.

RED BANEBERRY Buttercup Family

Actaea rubra

While this erect perennial is not a woody plant, its berries are so unusual and similar to fruits of shrubby plants that it is included here for comparison. In the shade of the open forest, this plant is 2 to 3 feet high. The compound leaves are divided into oval, toothed leaflets. The elongated flower cluster is 3 to 7 inches in length and consists of many small flowers, each with 4 to 10 small, white petals. There are two color phases to the fruits — red and white, prompting some botanists to classify these as separate varieties. The fragile berries are attractive to children, but there is strong evidence that they are poisonous.

Whitestem Gooseberry ¾ x

Rocky Mountain Maple ¾ x

Highbush Huckleberry ¾ x

Red Baneberry ¾ x

WOODLAND STRAWBERRY Rose Family

Fragaria vesca

This familiar wild flower and fruit inhabits the Lodgepole Pine forest. Even though it lacks erect stems, it produces horizontal runners which, in turn form new plants at their terminus. Such vegetative reproduction is very efficient. Each flower bears 5 green sepals, 5 rounded petals and numerous stamens. The pistils are numerous on a conical hump of tissue which becomes part of the edible fruit. Many of the wild strawberries have a more delicious flavor than the cultivated varieties. A great variety of wildlife enjoy the berries.

ARCTIC GENTIAN Gentian Family

Gentiana algida

Gentians are among our most esteemed wild flowers, being most abundant in mountains all around the world. The calyx of the flower is a cup or tube with 4 or 5 teeth; the corolla forms a tube or funnel with 4 or 5 lobes. Arctic Gentian is truly a dwarf alpine species found above 10,000 feet. It is only a few inches high, and is easily recognized by its greenish-white corolla which is spotted with dark purple. Blossoming occurs in August and September. *Gentiana* is named for Gentius, a king of Illyria who thought the plant had medicinal properties.

BUNCHBERRY DOGWOOD Dogwood Family

Cornus canadensis

This dwarf dogwood is a low growing plant with a horizontal undergound stem from which arise\ erect branches. These short branches are 3 to 8 inches high, crowned by 5 or 6 whorled leaves with sharp points at the tips. Above the whorl of leaves is borne a cluster of cream-colored flowers surrounded by 4 white, petal-like bracts which are often mistaken for petals. After shedding bracts and flower parts, the pistils produce cardinal red fruit clusters and each berry is about ¼ inch in diameter. This rare species is found in the open woods around Gibbon Meadow in Yellowstone; it blossoms in August.

Woodland Strawberry 1x

Arctic Gentian 1x

Bunchberry Dogwood ¾x

CASCADE MOUNTAIN ASH Rose Family

Sorbus scopulina

From the slopes of Teton Pass and Signal Mountain to Bechler River and Sylvan Lake, Cascade Mountain Ash is one of the most attractive shrubs of the Rose Family. The flat-topped flower clusters are 3 to 6 inches across. The brilliant red-orange fruits mature by late August and although they are bitter to human taste, they are useful to wildlife. Birds, especially grosbeaks and grouse, are regular feeders on the fruits, particularly in the winter time. The compound leaves are alternate and each one is composed of 5 to 13 leaflets, having finely toothed margins.

BIRCHLEAF SPIRAEA Rose Family

Spiraea betulifolia

In our western National Parks the spiraeas are mostly shrubs with showy clusters of tiny rose-like flowers. This particular species usually has a single erect stem between 1 and 3 feet high. The egg-shaped leaves are coarsely toothed towards the apex. The small, but perfect flowers have 5 sepals, 5 petals and numerous stamens. There are 5 distinct pistils which develop into pod-like fruits and open down one side at maturity. This is one of the most common shrubs of the two parks and is easily seen as one hikes the trails around the lakes. Insects frequently cause galls to form in the flowers thus detracting from their beauty. Many species of this genus are widely used as ornamentals.

Cascade Mountain-ash ¾ x

Cascade Mountain Ash ¾ x

Birchleaf Spirea ¾ x

COMMON CHOKECHERRY Rose Family

Prunus virginiana var. *melanocarpa*

The Common Chokecherry is a large shrub up to 20 feet in height and is known to many westerners as a source of small, tart cherries. The five petaled, white flowers are about ⅓ of an inch across and are borne in attractive cylindrical clusters of 3 to 4 inches long. The fruits, which gradually become bright red and finally almost black, ripen during August or early September. While tempting to the eye, the fruit is disappointing because it is harshly astringent to the taste and is nearly all stone. Many songbirds and mammals, however, relish the cherries and aid in the dispersal of seeds. The plant is common in the vicinity of Mammoth Hot Springs and along the banks of the Yellowstone and Snake Rivers. The fruits were very popular with the Indians; they would grind up the ripe fruit, stones and all, dry the material in the sun and store for later use. They also mixed this dried fruit with dried meat and fat to produce the famous concentrated food, pemmican. Chokecherries are suitable for making delicious jelly and the bark has been used to make a substitute for tea.

TUFTED EVENING PRIMROSE Evening Primrose Family

Oenothera caespitosa

This plant is also known as Moonrose, and is found on open, sunny slopes. Little or no stem is visible above the ground, but there are numerous toothed or pinnately cleft leaves. The flowers are composed of 4 sepals, 4 petals, 8 stamens, and 4 stigmas. The sweet scented blossoms are 2 to 2½ inches broad. When the flowers first open, they are white but turn pink as they mature. Some species in this genus have edible roots, leaves and shoots. The root is best when cooked in the spring.

Common Chokecherry ¾x

Common Chokecherry ¾x

Tufted Evening-primrose 1x

SMALL-FLOWERED WOODLANDSTAR Saxifrage Family

Lithophragma parviflorum

This small, herbaceous species has most of its leaves at the base of the stem with more or less lobed leaf blades. The flowers are in racemes with pink or white petals over 1/4 inch long. Note how each petal is deeply cleft into 2 to 4 narrow lobes, a feature that suggests the common name. Another species, *Lithophragma glabrum,* usually has bulbets in the axil of stem leaves. The flowering stage is found in June on grassy hillsides.

COMMON BEARGRASS Lily Family

Xerophyllum tenax

While limited to the southern part of Yellowstone and the northern part of Grand Teton, this species is very striking when it covers an open slope with flowering stems up to 5 feet tall. The base of the plant has a mass of wiry, grasslike leaves. The edges of the leaves are rough to the touch because of short, stiff hairs. The flowers are borne in a dense raceme which may be 6 to 8 inches long. The clumps of leaves arise from a thick rhizome and any particular offshoot may not flower for several years.

REDOSIER DOGWOOD Dogwood Family

Cornus stolonifera

The reddish-purple bark of the slender twigs of Redosier Dogwood makes the best field identification character. Combine this with the flat-topped clusters of flowers in the spring and white berries in the fall and it becomes the easiest shrub to recognize along the streams and rivers. The opposite leaves are lanceolate in shape and in September become some of the most brilliantly colored of the parks. Like other members of the family, this species bears flowers which are constructed on a plan of four — 4 sepals, 4 petals and 4 stamens. Seeds of the white to bluish fruits are eaten by many species of wild birds, especially the ruffed grouse. The foliage is browsed by mule deer, elk, moose and snowshoe hare.

Smallflower Woodlandstar 1x

Common Beargrass ⅛ x DO

Redosier Dogwood ¾ x

Redosier Dogwood ¾ x

RICHARDSON GERANIUM Geranium Family

Geranium richardsonii

Because of the long-beaked fruit, many species of this genus are known by the name of Cranesbill. Like its relative, *G. viscosissimum,* Richardson Geranium begins to bloom in June and continues into August. However, its habitat requirement is quite different. Instead of the sagebrush community or open woods, the Richardson Geranium grows in moist, Aspen woodlands often by a small stream. Sometimes freezing temperatures in mid-June will break down the leaf chlorophyll producing a premature fall coloration. The five parts of the ovary separate and are catapulted outward by the curving segments of the style.

WESTERN THIMBLEBERRY Rose Family

Rubus parviflorus

Thimbleberry belongs in the same group as the raspberries and blackberries, but can be readily separated from the latter by the undivided, simple leaves and the thornless stems. This wide-spreading shrub may reach from 2 to 6 feet high and is quite common along streams often in the shade of Aspen trees. The large, white flowers have a cylindric receptacle which is covered with numerous pistils. Following pollination and fertilization, this aggregation of pistils forms the bright red, raspberry-like fruit. These berries, which are quite tart, are relished by many kinds of birds and mammals. Because of the large, attractive flowers and the plant's ability to thrive in shady places, this shrub has good possibilities for planting around homes.

Richardson Geranium 1x

Western Thimbleberry ¾x

Western Thimbleberry ¾x

SMOOTH LABRADOR TEA Heath Family

Ledum glandulosum

While this small shrub is absent in Grand Teton, it can be found in a number of moist places in Yellowstone, such as Fire Hole River Canyon and Club Creek Canyon on the east entrance road. The oval to oblong leaves are fragrant when crushed and there is a tendency for the leaf margins to curl under. The white flowers are about ½ inch across, and the regular corolla is composed of five spreading petals. A similar eastern species was used as a substitute for tea, the use of which is responsible for the common name. The fruiting capsules are brown and split into five sections.

SHOWY GREEN-GENTIAN Gentian Family

Frasera speciosa

The Showy Green-gentian, one of the tallest of the herbaceous plants, has a thick stem with a curious mixture of numerous leaves and green flower whorls. Each individual flower has a plan of 4 sepals, 4 petals, and 4 stamens. Many insects are attracted to the rather elaborate nectar glands and purple spots on each petal. It can be seen along the Jackson Hole Highway or on Dunraven Pass during June and July. The first year the plant produces only a whorl of basal leaves, and in the second year the 2-5 foot unbranched stem appears.

SNOWBRUSH CEANOTHUS Buckthorn Family

Ceanothus velutinus

In California the species of *Ceanothus* are numerous and confusing, yet in Wyoming there are only three species. It is common around the shores of the piedmont lakes of Grand Teton and especially on the burned mountain slopes near Mammoth Hot Springs. Usually the plant is a rather low, rounded shrub 2 to 5 feet high. Another common name, Varnish Bush, comes from the shiny or somewhat sticky coating on the upper surface of the egg-shaped, leathery leaves. The small, white flowers are borne in showy clusters, and a close look at one flower will reveal the 5 petals have a scoop-like shape. A soapy lather may be obtained by rubbing a few flower clusters between the hands with a small amount of water. Deer, elk, and moose browse the herbage and birds eat the seeds.

SICKLETOP LOUSEWORT; PARROTS-BEAK Figwort Family

Pedicularis racemosa

The white flowers of Parrots-Beak distinguish it from the other two common species of this genus, the Bracted Lousewort and the Elephanthead. The curving and flattening of the 2 upper petals is responsible for the common name. In July the flowers are found along the Lakes Trail in Grand Teton, and in August it can be found at elevations as high as 9,500 ft. in the canyons or high passes.

Smooth Labrador Tea ¾x

Showy Green-gentian 1x

Snowbrush Ceanothus ¾x

Sickletop Lousewort ½x

BEARBERRY; KINNIKINICK Heath Family

Arctostaphylos uva-ursi

This trailing, evergreen shrub belongs to the group known as manzanita. It is the only manzanita found outside of western North America. Like other members of this group, Bearberry has the characteristic reddish-brown bark and smooth, leathery leaves. The white or pinkish flowers are bell-shaped and form in clusters at the ends of the stems. Later in the season the flowers give place to bright red berries which are a favorite food of bears and grouse. The common name, Kinnikinick, comes from an Indian expression referring to the use of the leaves and bark of this species with or in place of tobacco. Look for this shrub in sandy soil along the Snake River or in drier sites in Lodgepole Pine forests.

COMMON COWPARSNIP Parsley Family

Heracleum sphondylium

This hairy perennial herb is 3 to 8 feet tall and features umbellate flower clusters 4 to 6 inches broad. It is found from the valleys to moderate elevations in the mountains on moist ground. The stems and leaves are readily eaten by a variety of large animals including black bears. To render the stems palatable to humans, they can be peeled and boiled in two waters. However, no members of the Parsley Family should be eaten unless positive identification is possible.

Bearberry 2x

Bearberry ¾x

Common Cowparsnip ¼x

BLACK ELDERBERRY Honeysuckle Family

Sambucus racemosa

There are about 6 species of elderberries in the western United States, and they are all characterized by pithy stems, opposite, compound leaves and red to black berries. The plants are found along streams or moist, open forests. The small, numerous flowers are white to cream in color and arranged in flat-topped, compound clusters. By mid-August the solitary pistils form the fruits. Throughout the country these berries have been used for making jelly and wine and, although unpleasantly seedy, they are often used for pies. The roots, bark and mature leaves are considered poisonous to livestock; so use should be limited to fruits and flowers. The latter have been used to make a tea.

WESTERN SERVICEBERRY; SHADBUSH Rose Family

Amelanchier alnifolia

This small shrub may grow from 3 - 10 feet tall. The delicate, white flower clusters of serviceberry are among the first blooms of spring. Like other members of the Rose Family the flowers are built on a plan of five. The berries are ¼ to ⅜ inch in diameter, becoming dark purple at maturity. The juicy, sweet fruits are sought by many songbirds, but they are more important as food to the mammals. Squirrels, chipmunks and even bears seem to relish the apple-like berries. The mule deer, elk and moose are particularly fond of the foliage and twigs. Indians collected, dried and stored the fruits for winter food. They made a type of pemmican of pounded berries and dried meat to be carried on long trips. Authors of edible plant books recommend the berries for pies, jelly and wine.

Black Elderberry ¾ x

Black Elderberry ¾ x

Western Serviceberry 1x

Western Serviceberry ¾ x

WATER CROWFOOT; WHITE WATER BUTTERCUP Buttercup Family
Ranunculus aquatilis

The species of buttercup are so numerous that they offer some difficulty in identification. This aquatic species, however, is no problem because it is the only buttercup in the parks to have white flowers. All of the leaves are submerged, the blades being cut into numerous hair-like segments which mat together when extracted from the water. When not in flower this plant could be mistaken for Water Milfoil *(Myriophyllum).* Such aquatic plants in sluggish streams could favor the development of fish. The fruits of the flowers are achenes and are crowded onto a conical receptacle.

CRESTED WHEATGRASS Grass Family
Agropyron cristatum

Crested Wheatgrass has been widely introduced from the Old World to revegetate rangeland and dry pastures, and as a result has become well established in several parks. This tufted perennial grows from 20-40 inches tall, and like many other grasses has a very flattened, spike-like inflorescence. Individual florets have small green bracts, 3 stamens, and pollination is accomplished by the wind. There are over 4,000 species of grasses, and the family is one of the largest and most widespread. The value of grasses as a source of food for man and herbivores cannot be overestimated.

AMERICAN BISTORT Buckwheat Family
Polygonum bistortoides

A frequent herb of subalpine meadows, American Bistort has a flowering stem 12 to 24 inches high. If viewed from a distance, the crowded cluster of flowers has the appearance of a tuft of wool or cotton. The roots of this plant are starchy and quite edible either raw or boiled. Four miles north of Colter Bay on Highway 89 a very conspicuous meadow is filled with this species.

Water Crowfoot ¾ x

Crested Wheatgrass ¾ x

American Bistort 1x

FIELD CHICKWEED Pink Family

Cerastium arvense

Since the small, opposite leaves are covered with velvety hairs, they are suggestive of mouse-ears. The ½ inch broad flowers have petals which are deeply notched. The stem is usually less than 10 inches tall and tends to spread or sprawl on the ground. Close examination of the individual flowers reveal 10 minute stamens and 5 styles. Chickweeds as a group are often described as being weedy. The scientific name, *Cerastium,* comes from the Greek meaning "little horn" in reference to the shape of the fruit.

WESTERN VIRGINSBOWER Buttercup Family

Clematis ligusticifolia

This semi-woody vine has staminate and pistillate flowers on separate plants. The clusters of small flowers have no petals, but have 4-5 showy sepals. Later in the season the persistent styles become long and feathery. The vines grow over shrubs and trees along streams in the canyons. For sore throats and colds the Indians made an infusion of the plant; also, some tribes used the infusion for skin ailments such as eczema. This species is easily separated from *C. columbiana* by the clustered, small, white flowers.

HOT ROCK PENSTEMON Figwort Family

Penstemon deustus

Most Penstemons have blue to lavender flowers, but the species pictured is white to cream with conspicuous purple guide lines within the corolla tube (½ inch long). Note that the flower has bilateral symmetry with the upper lip consisting of 2 lobes and the lower lip with 3 lobes bent downward. The fifth stamen lacks an anther and is sterile. Most species in this large genus lack fragrant flowers, but this species has a mildly unpleasant odor. Look for this more or less woody plant in dry, open, often rocky places from 6,000 to 8,500 feet.

Field Chickweed 1x

Western Virginsbower 1x

Hot Rock Penstemon ¾x

CLASPLEAF TWISTED-STALK Lily Family

Streptopus amplexifolius

The flowers of this perennial herb are axillary, dangling at the ends of slender peduncles. The generic name, *Streptopus,* means "twisted foot" in reference to the bent or twisted flower stalks. The specific name, *amplexifolius,* means "clasping-leaf" in reference to the sessile leaves. This 2 to 4 foot plant grows only where the soil is wet along streams in the canyons. By August the greenish-white flowers have given way to bright red, oval berries about ½ inch long. The juicy berries may be eaten raw or cooked in stews or soup.

LADIES-TRESSES; PEARL TWIST Orchid Family

Spiranthes romanzoffiana

The creamy-white flowers of this plant are best appreciated with the aid of a hand lens. Like other orchid flowers, there are 3 sepals, 3 petals, and the stamens and pistil are combined into one unit. The small blossoms spiral up the slender 8-10 inch stem and appear in August and September. A few scattered plants are seen in the canyons, but most of them are found beside drying ponds at lower elevations.

WESTERN FALSE SOLOMON'S SEAL Lily Family

Smilacina racemosa

The erect and arching stems of this perennial herb arise from branching underground stems and may reach 2½ feet tall. The broad ovate leaves lack a petiole and tend to sheathe the stem. The numerous flowers are in a branched raceme. Like many others in the Lily Family, the perianth consists of 6 equal segments. This species is quite similar to Star Solomon's Seal, *Smilacina stellata,* but the latter plant has an unbranched raceme and relatively few flowers. Used in a judicious manner the bright red berries are edible. Flowering occurs in June and July.

Claspleaf Twisted-stalk ¾ x

Claspleaf Twisted-stalk 1 x

Ladies-tresses ¾ x

Western False Solomon's Seal ¾ x

TALL TOFIELDIA; FALSE ASPHODEL Lily Family

Tofieldia glutinosa

In Southern Europe the liliaceous genus *Asphodelus* occurs, and because of a superficial resemblance, Tall Tofieldia has been called False Asphodel. Perhaps a better common name might have been False Zygadenus since the plant looks much like Death Camas. Plant height varies from 6-18 inches and has grass-like basal leaves. The stem which emerges from a short, basal rootstock is usually speckled with purple glands. The species name, *glutinosa,* means sticky and refers to the glands on the stem. Wet, sandy soil around lakes or ponds provides the habitat requirements for this July and August bloomer.

WHITE BOG-ORCHID Orchid Family

Habenaria dilatata

Bog-orchid is frequently seen along stream banks or in bogs, thriving in full sun or in partial shade. The whiteness of the small flowers makes it easy to recognize. The lip, which has a dialated base, tapers to the tip. The spur is usually as long as the lip and projects outward from the rest of the flower. The flowers are fragrant with a pleasing spicy aroma. The flowers appear on the 10 to 20 inch stem in July and August.

NORTHERN BEDSTRAW Madder Family

Galium boreale

The minute, star-like flowers of Northern Bedstraw lack calyx members but have 4 spreading petals joined at the base. The fruit is composed of 2 round one-seeded pods which separate when they are ripe; these are necessary in identification of the species. The seeds are recommended as a substitute for coffee. The genus belongs in the same family as true coffee. The stems are square and the leaves are in whorls of 3 to 8. This plant thrives in open woods in damp soil.

COMMON TIMOTHY Grass Family

Phleum pratense

This species of the Grass Family is native of Eurasia and has repeatedly escaped from cultivation to become established in moist areas of our lower mountain valleys. This tufted perennial may reach 3 feet tall, the stems usually becoming enlarged and more or less bulbous at the base. The photograph reveals the flowering head at its most beautiful period. Each floret has 3 stamens, and at just the right time, the filaments elongate enough so that the delicate anthers are exposed for wind pollination.

Tall Tofieldia ¾ x

White Bog-orchid 2x

Northern Bedstraw ¾ x DO

Common Timothy 1x (Kret

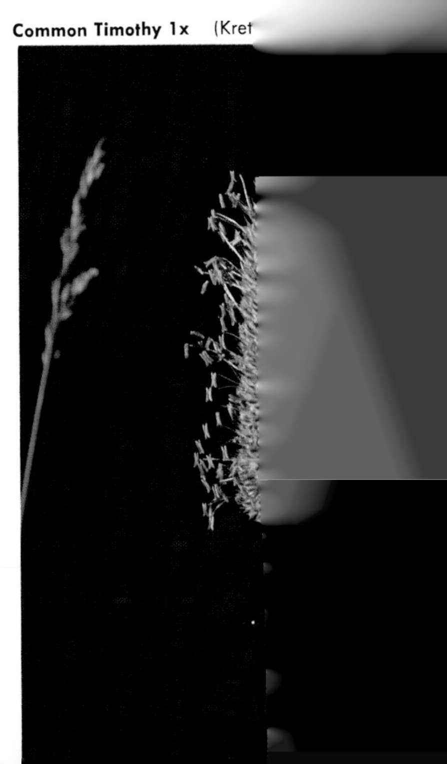

MANYFLOWERED PHLOX; FLOWERY PHLOX Phlox Family

Phlox multiflora

Driving the John D. Rockefeller, Jr. Memorial Parkway during June, one can see this low, mat-forming perennial. Open, wooded areas are best for the growth of this plant, but it often approaches timberline on such mountains as Mt. Washburn. The flowers of this genus are described as salverform, which means that the corolla has a definite tube crowned by lobes which extend at right angles. The leaves are very narrow and range between ½ to 1¼ inches long. The specific name *multiflora* means many flowers.

CANADA VIOLET Violet Family

Viola canadensis

This rather tall, perennial violet has wide distribution within the Rocky Mts. Recognition is relatively easy because of the white flowers and leaf blades which are often wider than long and broadly heart-shaped. The lower petal has several distinctive purple lines that are called guide lines by students of pollination. These lines are thought to be important in directing insect visitors past the stamens to the sweet nectar. Harrington suggests that all species of violets are edible either raw in salads or cooked as pot-herbs. Flowering occurs in late June or early July.

ROCKY MOUNTAIN PARNASSIA Saxifrage Family

Parnassia fimbriata

Around the lakes and along the streams of the canyons in late July and August this white-flowered Parnassia will surely catch your eye. The solitary flowers are on the ends of stems 2 to 12 inches high. The photograph highlights the daintily fringed bases of the petals which are so characteristic of this species. Note also that there are 5 white, fertile stamens alternating with 5 yellow, sterile stamens. The leaves are kidney or heart-shaped and have a smooth edge. This is a circumboreal species, occurring at high latitudes and extending southward along mountain chains.

Manyflowered Phlox 1x

Canada Violet 1x

Rocky Mountain Parnassia 1x

ALPINE SMELOWSKIA Mustard Family

Smelowskia calycina

Named after an early Russian botanist, Timotheus Smelowsky, this low herb is widespread in subalpine to alpine regions of eastern Asia and western North America. The leaves of Alpine Smelowskia form dense tufts or mats in rocky places; the petioles are conspicuously fringed with long hairs. The flowers are arranged in short, head-like racemes, becoming longer as the fruits develop. This fruit is about 4 times longer than it is wide. Flowering occurs in July and August. Climbers on the Grand Teton should watch for this well-adapted species.

COLORADO COLUMBINE Buttercup Family

Aquilegia caerulea

Few wild flowers have the delicate and ornamental grace of this member of the Buttercup Family. Throughout July and August it brightens the canyon trails up to 9,000 feet. The flowers usually have 5 petals which are extended backward into conspicuous, long, hollow spurs. The sepals vary in color from blue to white. The numerous yellow stamens and 5 long pistils project beyond the flower face. This columbine is the state flower of Colorado, and as one goes northward from Colorado, the blue color of the sepals fades to white or cream.

SEGO LILY; WIDEFRUIT MARIPOSA Lily Family

Calochortus eurycarpos

The genus *Calochortus* is one of the most beautiful in the Lily Family, and is readily recognized by the narrow sepals in contrast to the broad, conspicuously marked petals. Boiled bulbs have the flavor of potatoes and were eaten by the Indians and early settlers when food was scarce. However, the bulbs are hard to obtain because they are buried 4 to 12 inches below the surface, often in rocky soil. Several species of *Calochortus* once thrived in many western states, but cities and towns encroaching upon the habitat have left some species in a precarious position. June and July are the months for blossoms.

Alpine Smelowskia 1x

Colorado Columbine 1x

Sego Lily 1x (Despain)

WHITE DRYAS; EIGHT-PETAL DRYAD Rose Family

Dryas octopetala

Growing in limestone rocks and windy, exposed sites above 10,000 feet, this prostrate plant quickly catches one's eye because the flowers are so large (1½ inches across). One can readily see many adaptations to severe climatic conditions. For example, it has evergreen, leathery leaves with recurved margins to reduce water loss. It frequently grows with the tiny Arctic Willow. The 8 petals are responsible for the name *octopetala,* and the name *Dryas* comes from the Greek meaning "wood nymph."

MARSH MARIGOLD Buttercup Family

Caltha leptosepala

For the subalpine hiker the excitement of this plant comes with the discovery that blue buds of the flower push through the melting snow, and within 48 hours these blue buds expand into beautiful white blossoms, 1 to 2 inches across, similar to buttercups and anemones. The flowers are borne on leafless stems. The leaves are basal, succulent, and without lobes or divisions. The sepals are petaloid, and the petals are lacking. Some authors have recommended this plant as an edible potherb.

COMMON YAMPAH Carrot Family

Perideridia gairdneri

Yampah is a slender, erect herb from 1 to 3 feet tall. As an important food plant of the Indians and mountain men, it was recognized by its slender leaves and its 2 to 3 fleshy roots just below the ground level. When eaten raw, these roots have a carrot-like flavor, and can be ground into flour. The individual flowers are constructed on a plan of 5. They open in July or August depending on elevation. They are clustered in compound umbels, a feature common to all members of this family.

White Dryas 1x

Marsh Marigold 1x

Common Yampah ⅔x

ARUMLEAF ARROWHEAD Water-plantain Family

Sagittaria cuneata

The Arumleaf Arrowhead grows in the mud of shallow ponds throughout most of North America. There is variability in the general habit of the plants, but the plants always have long rootstocks that bear starchy tuber-like structures. The leaves are usually arrow-shaped and the flowers are in whorls of 3. Other common names are Swamp Potato, Tule Potato and Wapato. The Indian women's method of gathering the tubers was unique. They entered the water hanging onto a canoe and rooted out the tubers with their toes. The dislodged tubers rose to the surface and were placed in the canoe. The Indians would also seek out muskrat caches to obtain a supply. Roasted tubers taste better than raw tubers.

WHITE CAMPION Pink Family

Silene pratensis (Lychnis alba)

This European weedy perennial (or biennial) is widely distributed in North America. It has lanceolate leaves and varies between 2 to 4 feet in height. The male and female flowers are on separate plants. Those pictured are male flowers. The white flowers open at night when night-flying moths serve as pollinating agents. Close examination of the notched petals reveals that they have appendages forming a prominent circle at a point where the corolla emerges from the fused sepals. The female flowers mature into capsules which produce a great number of small seeds. This species inhabits roadsides and waste places.

COMMON PEARLYEVERLASTING Composite Family

Anaphalis margaritacea

This widely distributed perennial is covered with soft, woolly hairs especially on the lower surfaces of the lanceolate leaves. The small, white and yellow flower-heads are clustered into terminal, flat-topped arrangements. Each flower-head has a pearly-white series of bracts surrounding the inconspicuous flowers. The bracts and flowers have the ability to remain in good condition for weeks — long after one would expect the blossoms to fall off. Because of this ability, the name "everlasting" has been applied. Common Pearlyeverlasting is seen in August in dry, well-drained soils of the open forest.

Arumleaf Arrowhead ¾ x

White Campion 1½ x DO

Common Pearlyeverlasting 1x (Stockert)

ENGELMANN ASTER Composite Family

Aster engelmannii

Making the distinction between members of the genus *Aster* and the genus *Erigeron* is not easy. Many technical features are needed for identification. However this species is easily recognized by such reliable criteria as disk flowers surrounded by 15-19 white ray-flowers that may be 1 inch long. The leaves are lanceolate or elliptic and nearly smooth. The slightly hairy stems reach up to 3 feet high. This attractive plant can be found at 9,000 feet on Mt. Washburn or in the forest shade around Jenny Lake. The leaves may be boiled as greens.

COMMON YARROW Composite Family

Achillea millefolium

Before it blooms, this wild flower is sometimes mistaken for a fern because of its much divided leaves. The composite flowering heads are small and numerous with both ray and disk flowers. The scientific name *Achillea* is after Achilles who is supposed to have made an ointment from a relative of Yarrow to heal his wounded warriors after the siege of Troy. The Indians pulverized the plant and applied it to cuts, bruises and wounds. High in the canyons above 9,000 feet this plant may be only 6 to 8 inches tall; but in the valleys, it may reach 2 feet.

WHITEHEAD WYETHIA; WHITE MULE EARS Composite Family

Wyethia helianthoides

In the open, wet meadows north of Colter Bay and especially on Fountain Flats in Yellowstone one can find the most striking, white flowered composite. The broad, lanceolate leaves are responsible for the common name, and the generic name is in honor of Capt. N. J. Wyeth. The plant stands 2 feet tall and blends well with the Blue Camas of such wet meadows. The Mules-ear Wyethia, a closely related species, is often confused with Balsamroot. The flowering heads of this plant are so large they can be easily photographed without the aid of a portrait lens. The young plants are eaten by elk and deer.

Engelmann Aster 1x

Common Yarrow 1x

Whitehead Wyethia ⅓x

HAIRY GOLDEN ASTER Composite Family

Chrysopsis villosa

Species of this genus resemble the genus *Aster* in a number of ways, but differ in that the ray flowers are yellow instead of being white to purple. One feature, visible with a hand lens, is the presence of two whorls of pappus, the outer whorl being shorter than the inner whorl. Stems of the species illustrated grow 6 to 20 inches tall, bloom mostly in August and are common around the buildings at Old Faithful. *Villosa* means hairy, referring to the soft pubescence which covers the stem and leaves. This covering is not sticky, however, as it is in the Gumweed.

MULES-EAR WYETHIA Composite Family

Wyethia amplexicaulis

The bright green, shiny leaves of this plant are long and, in general, have the shape of a mule's ear; hence the common name. The generic name, *Wyethia,* is in honor of Capt. N. J. Wyeth, who crossed the continent with an early botanist in 1834. While the bright yellow heads are very similar to the sunflower, it is easily separated from the latter because the smooth stems are only 1 to 1½ feet tall. A closely related species, White Mule Ears, is found in wet meadows of both parks and hybridizes with *W. amplexicaulis*. Late June brings forth the flowers.

STEMLESS GOLDENWEED Composite Family

Haplopappus acaulis

There are many species of Goldenweeds in western North America, but the one illustrated is the most attractive, and usualy can be seen on the side road which parallels the Firehole River. The plant has a taproot and a crown of spreading branches. The individual composite heads have both ray and disk flowers whose intense yellow color makes them a worthy challenge to any photographer. The dense patches of growth plus the remnants of previous year's leaves should help to distinguish this plant from the Hairy Golden-aster or the Groundsel.

Hairy Goldenaster ¾x

Mules-ear Wyethia ½x

Stemless Goldenweed ¾x

NODDING BEGGARTICKS; BUR-MARIGOLD Composite Family
Bidens cernua

At the edges of ponds this freely branching herb will bloom in late July and August. The opposite leaves are sharply toothed and may join around the stem. The flower heads are usually 1½ inches across. The fruits are equipped with spines which catch on the fur of passing animals thus aiding in dispersing the species. Two distinct rows of green bracts are directly below the golden ray flowers. Some species of Beggarticks lack the ray flowers and are considered as unattractive weeds.

LARGEFLOWER HYMENOXYS Composite Family
Hymenoxys grandiflora

The almost perfect symmetry of the ray and disk flowers of this alpine species are in sharp contrast to the rocky, limestone slopes and ridges where it is found. The flower head is 2 to 3 inches in diameter, and the stem and leaves are covered with cottony hairs. Note how each ray flower has three lobes. The flowering heads always face towards the rising sun and remain facing eastward. The leaves are divided into slender segments.

LOW HAWKSBEARD Composite Family
Crepis modocensis

The genus *Crepis* is considered quite distinctive from other members of the family, but the separation of species in the genus is difficult because of the fact that many populations circumvent the normal sexual reproduction. In recognizing the hawksbeards, look for milky juice in stems and leaves, ray flowers only, leaves on stems and non-flattened seeds with a ring of white hairs at the top. The illustrated species grows in dry, open places, blooms in June and July, and has black hairs on the involucral bracts.

Nodding Beggarticks or Bur-Marigold 1x

Largeflower Hymenoxys 1x

Low Hawksbeard 1x (Stockert)

YELLOW SALSIFY; GOATSBEARD

Composite Family

Tragopogon dubius

This old-world species is an invader of waste places and roadside cuts. Its rapid spread in this country is due in part to the light, dandelion-like seeds which are carried great distances by the wind. The delicate fibers at the top of each seed act much like a parachute. These coarse herbs grow from thick biennial taproots, which in a related species, furnish the familiar salsify or vegetable oyster. When stems or leaves are broken, a milky juice is exuded. *Tragopogon* comes from two Greek words meaning "goat" and "beard," presumably referring to the conspicuous pappus at the top of the fruit.

ONE-FLOWER HELIANTHELLA

Composite Family

Helianthella uniflora

This species of *Helianthella* and its close relative, *H. quinquenervis,* can be easily confused with the true sunflower (*Helianthus annus*) and possibly the Showy Goldeneye (*Viguiera multiflora*). The leaves of *Helianthella* lack teeth, and the flower heads are borne singly, being 1½ to 2½ inches across. To separate the species illustrated here from *H. quinquenervis* check the back side of the leaves for prominent veins. *H. uniflora* has 3 and *H. quinquenervis* has 5. Look for the helianthellas in moist soil of open woods during July and August.

BALSAMROOT

Composite Family

Balsamorhiza sagittata

The June visitors frequently confuse this common plant with the sunflower, but careful examination reveals many striking differences. Numerous basal leaves are arrowhead shaped and covered with tiny, silvery hairs. When the Balsamroot is abundant along the highway, many folks comment about its very pungent odor. The stems are 1 to 2 feet tall, bearing solitary heads of yellow flowers at their terminus. It is located in open, sunny areas and seldom grows above 7,800 feet. The plants, especially the tender shoots, are eaten by deer and elk.

Yellow Salsify ¾x

One-flower Helianthella ¼x

Arrowleaf Balsamroot ¼x (Condon)

COMMON TANSY

Composite Family

Tanacetum vulgare

Common Tansy is an Eurasian species which was cultivated in North America and escaped. It was used medicinally by the Indians after introduction. The Europeans officially used tansy to induce menstruation, and yet the Indians utilized a tea made from the whole plant to induce abortion (sometimes with disastrous results). The stem length varies from 2-5 feet and is covered with finely dissected leaves. Look for it in disturbance sites during August and September.

CURLYCUP GUMWEED; RESINWEED

Composite Family

Grindelia squarrosa

Curlycup Gumweed is noted for the abundant sticky resin which it exudes especially from the outward curving involucral bracts. The stems grow from 10 to 25 inches high and the numerous flower heads are about 1½ inches across when fully mature. Indians of northern California pounded the resinous flower heads and applied the resulting gum to relieve poison ivy inflamations. They also used this gum for the treatment of asthma, bronchitis, and whooping cough. This is one of the first species to invade disturbed sites such as roadside cuts. Flowering begins in August and may continue through September.

SHRUBBY GOLDENWEED

Composite Family

Haplopappus suffruiticosus

Members of the genus *Haplopappus* are widespread in the western United States, but they are a difficult group to separate into species. The illustrated species is easy though, because it is found in rocky places at high elevations such as below the summit of Mt. Washburn. (Y.N.P.). The slightly woody stems are 12 to 20 inches high and bear showy flower heads with 3 to 6 ray flowers. The leaves are pleasantly fragrant, have glandular hairs, and a wavy or crisped surface. August is the month for blossoms.

Common Tansy 1x

Curlycup Gumweed 1x

Shrubby Goldenweed 1x

COMMON RABBITBRUSH Composite Family

Chrysothamnus nauseosus

Rabbitbrush plants inhabit the arid sagebrush areas of both parks. During most of the summer, these plants are unattractive, but in late August the bushes become covered with numerous heads of golden-yellow flowers. Many insect pollinators are attracted to these masses of flowers even though the individual heads lack the ray flowers so common in other members of the family. Besides furnishing useful cover in open areas, rabbitbrush is important to wildlife as the foliage and seeds are readily consumed, especially by rabbits and hoofed browsers. This species is 2 to 3 feet high and both stems and leaves are covered with a white, woolly pubescense.

HEARTLEAF ARNICA Composite Family

Arnica cordifolia

The common name of this plant arises from the fact that basal and lower stem leaves are markedly heart-shaped at the base. The medicine bearing this same name (Arnica) is obtained from *Arnica montana,* a European species. A plant of open forests, it has large, composite flowering heads composed of light golden ray flowers and numerous, deep yellow disk flowers. Blossoming begins in mid-June and lasts until mid-July. A few individuals start to bloom again in early September.

PRAIRIE CONEFLOWER Composite Family

Ratibida columnifera

This species is a recent newcomer to the two parks, but could become a permanent resident along the major highways. The bright yellow ray flowers immediatey set it apart from the more widespread Western Coneflower of the genus *Rudbeckia.* It is usually the middle of August before one is aware of the long, thimble-like centers, but careful observation of the tiny, numerous disk flowers reveals that all flower parts are present.

WOOLLY ERIOPHYLLUM Composite Family

Eriophyllum lanatum

On the sagebrush flats and moraines one is sure to see this yellow composite which has leaves and buds covered with white hairs. Additional common names include Woolly Yellow Daisy and Yellow Woolly Aster. Even though it is not a true daisy nor aster, it superficially resembles the previously mentioned species. The plants form conspicuous clumps from 10 to 18 inches high, covered with bright, golden, composite heads nearly an inch in diameter. It blooms during July and August. *Eriophyllum* refers to the woolly character of the leaves.

Common Rabbitbrush ½x

Heartleaf Arnica 1x

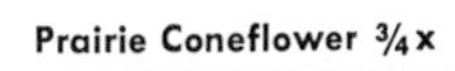

Prairie Coneflower ¾x

Woolly Eriophyllum ¾x

MISSOURI GOLDENROD Composite Family

Solidago missouriensis

It is difficult to differentiate the many species of goldenrods found in North America. They are all erect perennials, bearing alternate leaves and small composite heads containing both disk and ray flowers. The species illustrated is between 1 and 2½ feet high and has basal leaves 2 to 7 inches long. The leaves on the stem are very narrow and have a smooth margin. The many, small flower-heads are arranged on only one side of the spreading branches. The flowering period is in August and September, and this species grows in a variety of dry, open sites.

SHOWY GOLDENEYE Composite Family

Viguiera multiflora

Long after the spring and summer flowers have faded from the mountain landscape, many roadsides are brightened by the golden-yellow heads of the Showy Goldeneye. Like the Heartleaf Arnica and the Balsamroot, this plant has composite flowering heads composed of light golden, ray flowers and numerous, yellow disk flowers. The heads are few, with from 10 to 14 rays about 1 inch long. This perennial herb has willow-like leaves which are oppositely arranged at least in the lowermost part.

SNOW CINQUEFOIL Rose Family

Potentilla nivea

This dwarf perennial species may be found on rocky slopes and high altitude moraines. The most conspicuous alpine adaptation displayed by the plant is the long, soft hairs covering the stems and leaves. These hairs help to reduce evaporation of water and the intensity and directness of strong alpine light. *Potentilla* is a large and taxonomically difficult genus of at least 200 species. The genus might be mistaken for the genera *Rosa* or *Ranunculus* (buttercups), but 5 bracts below and alternating with the 5 sepals make for a quick distinction.

Missouri Goldenrod ¾x

Showy Goldeneye 1x

Snow Cinquefoil 1x DO

BIG SAGEBRUSH Composite Family

Artemisia tridentata

This fragrant, grayish-green shrub is well represented in both parks, especially on the valley floor of Jackson Hole. Here it is dominant, and the foliage and flower clusters constitute most of the diet of the Sage Grouse. Antelope and deer also make good use of sagebrush as forage. In the parks this shrub may vary from 1 to 5 feet high and have a definite trunk up to 3 inches in diameter. The yellowish flowers are borne in small heads and do not appear until late August or early September.

MOUNTAIN DANDELION; PALE AGOSERIS Composite Family

Agoseris glauca

Like the Common Dandelion this species has a milky juice in stems and leaves, only ray flowers and a basal rosette of leaves. Technical differences of the fruits, seen only with a hand lens, separate the two groups. The flowering stems, which are from 6 to 20 inches in height, have no leaves. The flower heads are up to 1½ inches broad and are attractive to a variety of pollinators. The fruits are distributed by the wind catching the many fine bristles which surround a well-defined beak. Habitats vary from rocky sagebrush conditions to subalpine meadows.

WESTERN GROUNDSEL Composite Family

Senecio integerrimus

As a genus, *Senecio* is one of the largest genera, having at least 1,000 species described. At least 12 species occur in both parks. These species have yellow to yellow-orange flowers in heads bearing both ray and disk flowers. Senecios might be confused with arnicas, but the latter species have lower leaves that are opposite while groundsels have alternate leaves. Western Groundsel stems vary from 8 to 24 inches tall and when young are covered with soft, silky hairs. The groundsel illustrated here blooms in June and July and requires a sandy to gravelly soil.

EVENING PRIMROSE Evening Primrose Family

Oenothera hookeri

This is a difficult genus in which species identification can be frustrating. All the species have an inferior ovary and a long hypanthium (fused bases of sepals, petals and stamens) which serve to elevate the showy petals, stamens and stigma above the lanceolate leaves. The anthers are balanced on the filaments at their mid-point and thus are able to swing to and fro in the wind. The tall style is topped by a stigma that is divided into 4 finger-like lobes which spread star-like. This species inhabits moist sites and especially roadsides.

Big Sagebrush

Mountain Dandelion 1x (Stockert)

Western Groundsel 1x (Stockert)

Evening Primrose ½x DO

SHRUBBY CINQUEFOIL Rose Family

Potentilla fruticosa

Beginning in June and extending into September, this shrub has one of the longest blooming periods of any species in the parks. It is common along the Snake River and many accessible places in Yellowstone, such as Gibbon Meadow and Mount Washburn. Since the flower has 5 petals, 5 sepals, and numerous stamens, it is very much like a wild rose, but the plant is always thornless. The leafy stems are erect or ascending and grow to 4 feet in height. The generic name, *Potentilla,* is from the Latin word *poten* referring to the "powerful" medicinal value of some species. The species name, *fruticosa,* means "shrubby." This shrub is worthy of cultivation and many horticultural varieties are available through nurserymen.

YELLOW MONKEY-FLOWER Figwort Family

Mimulus guttatus

Growing in wet meadows and along stream banks, the Yellow Monkey-Flower's bright, yellow, spotted petals draw immediate attention. Delicate hairs cover the three lower lobes of the corolla and, together with the orange spots, help to attract insect pollinators. Close examination of the stigma reveals two roundish lobes which are spread apart. When one of these lobes makes contact with a pollen-laden bee, the two stigma lobes immediately begin to come together like the leaves of a book. The pollen will thus be held firmly and when the bee backs out of the flower, no self-pollination will occur. The square stems have opposite leaves and are quite fragile because of their hollow structure.

LANCELEAVED STONECROP Stonecrop Family

Sedum lanceolatum

Recognition of stonecrops is easy because of the succulent nature of their leaves and stems, but the separation of species has been confusing, calling for careful attention to details. Lanceleaved Stonecrop has numerous basal rosettes of leaves. The leaves on the stem vary greatly in shape and they are not ridged underneath. The numerous flowers are clustered at the summit of a 2 to 9 inch stem. Each flower has 4 or 5 narrow, jointed petals and 8 to 10 stamens. The fruit pods (follicles) stand erect. Look for this species on rocks or gravelly soil. Flowers occur from late June through August.

Shrubby Cinquefoil 1x

Common Monkeyflower 1x

Lanceleaf Stonecrop 1x DO

UTAH HONEYSUCKLE Honeysuckle Family

Lonicera utahensis

Known also as Red Twinberry, this shrub of the Lodgepole Pine forest varies from 2 to 5 feet high. In June the paired, yellow flowers are a delight to behold. The five petals are united into a funnel-shaped tube which is slightly swollen at the base. The fruits are attached to the axil of the leaf by short, 1 inch pedicels. Flowers and fruits separate this species from the Bearberry Honeysuckle because they lack the subtending involucral bracts. The fruits are eaten by birds and chipmunks, and while some reports tell of a poisonous substance being present, there is little danger to children because the fruits are not palatable.

BEARBERRY HONEYSUCKLE Honeysuckle Family

Lonicera involucrata

Although there are about 24 native species of honeysuckle in the United States, there are only 3 species in the two parks. They are often confused, but they can be readily separated by checking the base of the flower or fruit. Bearberry Honeysuckle always has conspicuous involucral bracts which subtend these structures. This branching shrub stands 1 to 6 feet high and has four-angled twigs. The yellow flowers are aways attached in pairs in the axils of the leaves. Following the June flowers, the fruits begin their development and by late August these berries are juicy and purple-black in appearance. The persistent involucral bracts become dark red and expand to expose the berries.

Utah Honeysuckle 2x

Utah Honeysuckle 1x

Bearberry Honeysuckle ¾ x DO

Bearberry Honeysuckle ¾ x

MOUNTAIN GOLDENPEA Pea Family

Thermopsis montana

Because of a superficial resemblance, this plant is sometimes called False Lupine, but distinctive features will set it apart easily. The leaves are trifoliately compound with leaflets up to 4 inches long, and the 10 stamens are always separate and distinct. As the petals fall off, the ovary elongates and at maturity the flat fruits may be erect or horizontal. Goldenpea is unpalatable to game and grazing livestock so it may replace more desirable forage plants. It inhabits wet meadows and blooms in July and August.

BUTTER-AND-EGGS Figwort Family

Linaria vulgaris

Throughout North America the Butter-and-eggs plant has made itself a familiar habitant of the roadsides and waste places. It closely resembles the cultivated snapdragon except that the inch-long corolla is spurred at the base. Slender stems, 1 to 2 feet tall, and bearing numerous, narrowly linear leaves, arise from perennial roots. A native of Europe, this plant has often escaped cultivation and formed large patches from the creeping roots. The cylindrical capsules produce winged seeds.

DALMATIAN TOADFLAX Figwort Family

Linaria dalmatica

Dalmatian Toadflax is a native of Southeast Europe. It was brought to North America as an ornamental but has escaped and become weedy. It is common in the area of Mammoth Hot Springs and occasionally is seen along roadsides in Grand Teton. The plant has flowers similar to the Butter-and-eggs plant, but has ovate, clasping, linear leaves. Both species have numerous seeds and spread by creeping roots, features which make control difficult.

Mountain Goldenpea ¾ x

Butter-and-eggs ¾ x

Dalmatian Toadflax ¾ x

GLACIER LILY Lily Family

Erythronium grandiflorum

Common names of plants frequently bring about confusion among flower lovers because they vary so much in different geographical areas. For example, this plant is also known by such names as Dogtooth Violet, Adders Tongue, Fawn Lily, and Trout Lily. In both parks the plants are abundant above 7,500 feet. The yellow, nodding flowers are on a stem 6 to 12 inches high, and this stem arises from a bulb several inches below the soil surface. The flowers open early in the season while the snow is still melting.

PLAINS PRICKLYPEAR Cactus Family

Opuntia polycantha

The Cactus Family is found mostly in the New World, and while our parks have only 2 species, parks of the southwest may have dozens of species. The species illustrated here is frequently seen in the Mammoth area; *O. fragilis* is seen on gravelly terraces above the Snake River. The cacti can be recognized by thick, fleshy, green stems which are armed with spines. The conspicuous, yellow flowers possess numerous sepals, petals and stamens. The Blackfeet Indians treated warts by lacerating them and applying the fuzz from this cactus.

COMMON BLADDERWORT Bladderwort Family

Utricularia vulgaris

This circumboreal species has submerged, floating stems and leaves. The leaves are divided into thin, linear segments which bear small bladders of unique construction. Microscopic, aquatic animals are trapped and supposedly digested by these bladders. From the submerged stem rises an erect flowering stem up to 4 inches tall. Note the similarity of the yellow flowers to those of Butter-and-eggs. Look for the flowers in quiet water during July and August.

Glacier Lily ¾x

Plains Pricklypear ¾x (Stockert)

Common Bladderwort 1x

BLAZING STAR MENTZELIA Blazing Star Family
Mentzelia laevicaulis

One reason that this plant draws such frequent comments is that it occupies unlikely sites, such as gravelly road side cuts and dry streambeds, where many plants fail to invade. Because barb-like hairs on the leaves adhere to clothing and to the hair of animals, these plants are sometimes called Stickleaf. The light yellow flowers which are 1 to 3 inches in diameter are borne in a branching inflorescence at the end of 2 to 3 ft. stems. The many stamens are nearly as long as the petals but tend to clump together. The Indians parched and ground the oily seeds to make a nutritious flour.

COMMON ST. JOHNWORT St. Johnwort Family
Hypericum perforatum

This Old-World species is a well-established perennial weed over much of eastern and western United States. In the western states it has acquired the name Klamath-weed and has become a serious pest in pastures. It has established itself in the two parks since 1971 and will undoubtedly spread to any new disturbance sites. It is much taller than *H. formosum* reaching up to 3 feet high. Even though this plant is poisonous to livestock, it has been used medicinally by the Indians and Europeans for many ailments from tuberculosis to stones in the bladder.

WESTERN WALLFLOWER; PRAIRIE ROCKET Mustard Family
Erysimum asperum

Within both parks there are many species of wild mustards, but this is the most attractive. It is infrequent in dry, stony places, and the bright yellow petals are ½ inch long. The flowers are clustered into a round, terminal head, but as the plant matures, narrow, mostly four-sided pods develop reaching a length of 2-4 inches. The stem leaves are narrow and sometimes have small teeth. There is confusion over the correct scientific name of this species. As a group the genus needs more study.

Blazing Star Mentzelia (Courtesy of G.T.N.P.)

Common St. Johnwort 1x

Western Wallflower ¾x

SUBALPINE BUTTERCUP Buttercup Family

Ranunculus eschscholtzii

The waxy, five-petaled flowers of this talus slope perennial typify the some 12 or more Buttercup species growing in the parks. The plant has one to several stems, usually with yellow or brownish hairs below the flowers. There are 3 varieties of this species and these are separated on the variation of the basal leaves. They vary from shallowly 3-lobed to divided into segments. The numerous species of *Ranunculus* offer considerable difficulty in identification since some will hybridize, leading to intermediate forms. The features of the fruits are the most useful.

AMERICAN GLOBEFLOWER Buttercup Family

Trollius laxus

This smooth, perennial herb bears its large, conspicuous flowers singly at the ends of its several stems. There are no petals, but the 5 to 8 cream-colored sepals are often ½ inch long. *Trollius* might be confused with *Caltha leptosepala* but the former has deeply cleft or divided leaves and the latter lacks the lobed leaves. The fruits in this species are follicles with several seeds. These plants inhabit swampy ground to alpine meadows generally blossoming near melting snow.

CLIFF ANEMONE; WINDFLOWER Buttercup Family

Anemone multifida

Members of the genus *Anemone* lack petals but have colorful, petal-like sepals. The stamens and pistils are numerous; the latter parts develop into small, dry, one-seeded fruits called achenes. In both parks this species has considerable variation in the color of sepals — from yellow to purple. Habitat sites include wet areas along the Snake River to alpine sites above 10,000 feet. The principal leaves are basal and are palmately lobed or divided. Additional leaves are attached higher on the stem.

Subalpine Buttercup 1x DO

American Globeflower 1x

Cliff Anemone 1x

MEADOW DEATHCAMAS; DEADLY ZIGADENUS Lily Family

Zigadenus venenosus

Deathcamas is the common name for several species of this genus that are poisonous to livestock and man. Toxic alkaloids occur throughout the plants, and animals are poisoned by leaves, stems and flowers. Bulbs of this species were occasionally confused with Blue Camas by Indians and early settlers and caused disastrous results. The stems of this perennial grow from 1 to 2 feet tall when in full flower. The individual blossoms are about ¼ inch wide and have the typical lily flower constructed on a plan of 3. In both parks look for the flowers in June and July in sagebrush communities.

SULFUR BUCKWHEAT Buckwheat Family

Eriogonum umbellatum

Also known as Sulfurflower, this plant is especially showy in the rocky, sagebrush flats. The sulphur-yellow blossoms are in an umbrella-like cluster at the top of a 10 to 12 inch flowering stalk, while the numerous leaves grow very near to the ground and accumulate bits of organic matter which eventually become part of the soil. A closely related alpine species (*E. ovalifolium*) has pale yellow flowers which become tinged with pink in drying. The flowering period extends from mid-June to early August.

CREEPING BARBERRY; OREGONGRAPE Barberry Family

Berberis repens

Creeping Barberry is a common, low shrub less than 18 inches high. It is most frequently found on well-drained, morainal soil. Every spring dense clusters of yellow flowers brighten the appearance of prickly, evergreen leaves. Usually each flower has 6 sepals, 6 petals and 6 stamens. By August the plants have produced rather sparse clusters of grape-like fruit. Along with the ripening of the fruit, the leaves turn beautiful shades of red or purple. Some people claim the berries have a bitter taste, but a number of authors recommend using them in making jelly, jam or wine.

Meadow Deathcamas 1x (Stockert)

Creeping Barberry 2x

Sulfur Buckwheat 1½x

Creeping Barberry ¾x

NUTTALL VIOLET Violet Family

Viola nuttallii

There are two species of yellow violets which bloom during the first two weeks of June. The one pictured is the larger of the two and has the larger leaf. All violets have irregular flowers, consisting of 5 sepals, 5 petals and 5 stamens. The lowest petal bears a sac-like spur at its base and contains nectar. The whole flower arrangement favors cross fertilization. The genus *Viola* is considered a critical group with many difficult species, primarily because they hybridize freely under natural conditions.

ROCKY MOUNTAIN PONDLILY; SPATTERDOCK Waterlily Family

Nuphar polysepalum

This large, attractive, aquatic plant is also known as Cowlily and can often be found in great numbers covering the surfaces of beaver ponds or lakes. The showy, yellow flowers are from 2 to 3 inches across, having 5 to 6 sepals and many stamen-like petals. The shiny, green leaves are 5 to 12 inches long and 5 to 9 inches broad. The plant's rootstalks are buried in the mud. These rootstalks are said to be stored by muskrats in their burrows. The Indians raided these caches for emergency food.

BRACTED LOUSEWORT; FERNLEAF Figwort Family

Pedicularis bracteosa

The repellent common name, lousewort, comes directly from the scientific Latin name, because many years ago farmers believed that when their cattle fed upon the flowers of this plant, the animals might become infested with tiny lice called pediculus. The individual flowers are ½ to ¾ inch long and are curved downward at the tip. It is found at nearly all elevations from 6,800 feet to 8,500 feet in late June and July.

YELLOW COLUMBINE Buttercup Family

Aquilegia flavescens

The 5 long tubes or "spurs" of the Columbine flower are really petals. There are 5 pistils and many stamens. The scientific name seems to be derived from the Latin *aquila* meaning "eagle"; apparently the spurred petals suggest an eagle's talons. The leaves are divided into many small segments. These plants inhabit streams and some roadside cuts. Below Isa Lake on the road to Old Faithful, there is a population of these beautiful Yellow Columbines.

Nutall Violet 2x

Rocky Mountain Pondlily ¼ x

Bracted Lousewort ¾ x

Yellow Columbine ½ x

YELLOWBELL; YELLOW FRITILLARY Lily Family

Fritillaria pudica

The underground bulb of this plant contains starch and is edible either raw or cooked. As the bulb develops it becomes surrounded by dozens of small bulblets, each of which may become a new plant. The flowering stem appears in late May or early June in sagebrush habitats. The yellow or orange drooping flowers turn reddish as they age. The perianth parts are about ¾ inch long and when the perianth parts fall and the fruit starts to ripen, the stem straightens out placing the three-sectioned fruit in an erect position.

WESTERN ST. JOHNWORT St. Johnwort Family

Hypericum formosum

During the middle ages the European herbalists developed many superstitions and medicinal properties in connection with the Common St. Johnwort of Europe which later migrated to America. There are several American species which are usually found between 8,000 and 10,000 ft. The species at the right has an elongated, open inflorescence, and the unopened flower buds are conspicuously red in contrast to the yellow mature flowers. Under a hand lens, the leaves show small, black dots on the margins.

STONESEED; COLUMBIA PUCCOON Borage Family

Lithospermum ruderale

Shoshoni women reportedly drank an infusion of Stoneseed root every day to act as a contraceptive. Experiments using alcoholic extracts of the plant on mice eliminated the estrous cycle, thus verifying in part the Indian use. This perennial has a clump of hairy stems and lanceolate leaves, 1 to 4 inches long. The flowers are in small clusters in the upper axils of the leaves. The plant grows in relatively dry places up to mid-elevations in the mountains. The generic name, *Lithospermum,* comes from two Greek words, stone and seed, a reference to the very hard stony seeds. The common name, Puccoon, is an Indian word for plants yielding dyes.

BUTTERWEED GROUNDSEL Composite Family

Senecio serra

In the northern part of Yellowstone and the southern part of Grand Teton this tallest species of Groundsel will bloom in profusion during August. The stems can be up to 40 inches high, and the narrow leaves are sessile and toothed. Its nearest look-alike, *S. triangularis,* has lower leaves that are triangular in shape. This species thrives well under aspen and occasionally on dry, open slopes. This stout perennial herb often spreads from short rhizomes.

Yellowbell 1x

Western St. Johnwort 1¼x DO

Stoneseed 1x DO

Butterweed Groundsel ½x

FIVE-STAMEN MITERWORT

Saxifrage Family

Mitella pentandra

The flowers of miterworts are famous for their petals, which are cut or divided into narrow lobes or segments. The species illustrated is best appreciated by using a hand lens on the flower. Under magnification it will be noted that each stamen stands opposite a petal. The leaves are roundish and have a toothed margin. The flower stems stand from 10 to 15 inches tall. July and August are the months for flowers, and the plants thrive in shaded, moist woods of the canyons. The common name, miterwort, derives from the two-beaked capsule, which has the appearance of a bishop's miter.

SULFUR PAINTBRUSH

Figwort Family

Castilleja sulphurea

The paintbrushes are a striking and beautiful group which are difficult to separate into species, mainly because the inconspicuous flowers are surrounded by brightly colored bracts. These bracts resemble leaves except for their color. The calyx has 4 lobes and the greenish corolla is designed with a hooded upper lip and a three-toothed lower lip. In some meadows a perplexing mixture of bract colors occurs, indicating hybridization between species is possible. The species illustrated will be found in moist meadows up to 11,000 feet throughout the Rocky Mountains. *Sulphurea* means sulphur-colored referring to the bracts.

ANTELOPE BITTERBRUSH

Rose Family

Purshia tridentata

This partly prostrate shrub reaches 3 feet in height and usually grows in association with sagebrush. It is common among the flats of the Madison River and the terraces above the Snake River. The leaves are small but 3-toothed at the tip. The yellow flowers have 5 petals, 5 sepals and, numerous stamens. The foiliage and twigs are bitter as the common name implies, but in spite of this, it is a very valuable winter and spring browse for elk and deer. The spindleshaped fruits are quite pubescent and provide food for small mammals like chipmunks and ground squirrels. The Indians used to brew the leaves to prepare a cough medicine.

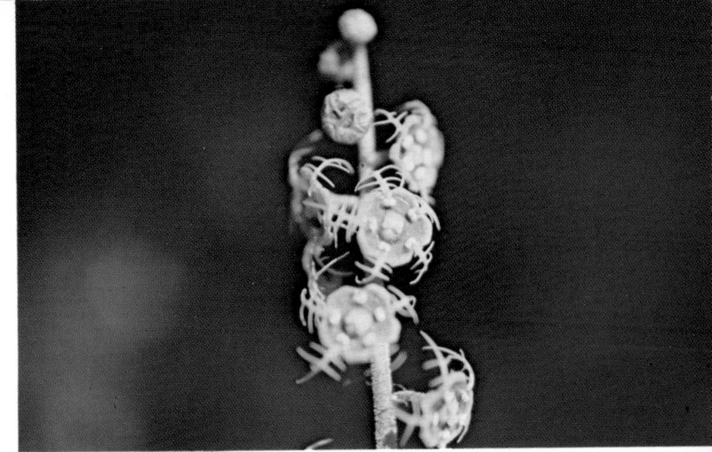

Five-stamen Miterwort 2x

Sulfur Paintbrush 1x

Antelope Bitterbrush ¾x

SPRINGBEAUTY Purslane Family

Claytonia lanceolata

Springbeauties are widespread throughout both parks and are among the first flowers to follow the retreating snowbanks. Several stems grow from a tuberous underground stem. Each of these stems has two basal leaves which are rather fleshy. Flower color varies from white to pink, and in the whiter forms pinkish veins add emphasis to the notched petals. Some plants are found even above 9,000 feet during August. The tuberous stems (½ to ¾ inch in diameter) were dug by the Indians and eaten as we would eat potatoes.

STICKY GERANIUM Geranium Family

Geranium viscosissimum

Also known in many localities as Cranesbill, this species is a common late spring flower which continues blooming into late August. Found in sagebrush and open woods, the plant grows to a height of one or 2 feet. The rose-purple flowers (1 to 1½ inches across) are constructed on a plan of five: i.e., all parts are in five or multiples of five. The fruit, which resembles a crane's bill, is unique in its method of seed dispersal; as the capsule ripens, its longitudinal sections split open with such recoiling force that the seeds are catapulted outward from the parent plant for several feet.

NORTHERN TWINFLOWER Honeysuckle Family

Linnaea borealis

This creeping evergreen herb is found in the shade of the coniferous forest. From the trailing branches rise flower stalks which bear two pendent pink flowers. The corolla is almost equally five lobed, but there are only four stamens. The generic name of the plant is named for Carl Linnaeus, the Swedish botanist. This species is found around the world in northern latitudes. These plants are eaten by grouse and deer.

Springbeauty 1x

Sticky Geranium 1x

Northern Twinflower 1x

PARRY PRIMROSE Primrose Family

Primula parryi

One of the largest, subalpine, herbaceous plants of Grand Teton National Park stands out conspicuously because of the magenta flowers. Several flowers, each on a nodding pedicel, are clustered at the top of a stout stem. Close examination of a flower reveals that the 5 petal lobes join at their base into a narrow tube. These flowers have a disagreeable odor, which undoubtedly, attracts some bizarre insect pollinator. In the middle of a rushing stream or on a mossy ledge, one is apt to find this spectacular primrose. During August look for this plant in Cascade and Indian Paintbrush Canyons above 9,000 feet.

WILD ROSE; WOODS ROSE Rose Family

Rosa woodsii

The wild rose is perhaps the most quickly recognized of all the flowering shrubs. Having an almost universal distribution in the United States, people remember its characteristic large flowers, compound leaves and peculiar fruits called rose hips. The showy flower, in addition to numerous stamens, has several pistils which are enclosed by the cup-like receptacle. After the petals fall, this entire structure becomes an aggregate fruit of great importance to wildlife. The hips remain on the shrubs throughout the winter providing food for large birds and hoofed browsers. Early westerners made jelly from the ripe fruit and ate hips raw from the bush. The Indians also gathered the fruits for food and used the roots for treatment of ailments.

Parry Primrose 1x

Wild Rose ¾x

Wild Rose ¾x

PRAIRIESMOKE Rose Family

Geum triflorum

The silvery tails of the fruit and the nodding flowers of this plant have been responsible for many common names, such as Grandfathers Beard, Long-plummed Avens and China Bells. The stems are from 7 to 20 inches high and have mostly basal and fernlike leaves. The fruits of this plant are dispersed by the wind's catching the long, feathery styles. Look for this hairy plant in open meadows, hillsides and ridges up to about 8,000 feet.

PINK WINTERGREEN; SHINLEAF Heath Family

Pyrola asarifolia

Wet soil around streams and springs and the shade of coniferous forests are places to look for this evergreen perennial. The flowers are arranged in slender racemes on stems 8 to 15 inches high. There are 5 petals and 10 stamens, the anthers of which release their pollen through terminal pores. The style is characteristically bent to one side and often has a ring or collar below the stigma. The common name, Shinleaf, comes from the early use of the leaves in making plasters for injured shins.

HOLBOELL ROCKCRESS Mustard Family

Arabis holboelii

Identification of genera and species in the Mustard Family is difficult or nearly impossible without mature fruits. Particularly in this genus, it is necessary to know if the pods (siliques) point up, down or sideways. The flowers consist of 4 sepals, 4 white to pink petals, 6 stamens and 2 united pistils. The basal leaves are broadest at or near the tip. Flowering specimens can be found from June until mid-August in the sagebrush or open forest.

BITTERROOT LEWISIA Purslane Family

Lewisia rediviva

It is fitting that a plant with such startling beauty and historical importance was chosen as the state flower of Montana. The starchy roots were eagerly sought by the Indians as a source of food, and they introduced the species to Captain Lewis of the Lewis and Clark Expedition. Later a British botanist named the plant *Lewisia,* after the explorer. The fleshy leaves appear as soon as the snow melts and usually dry up at the time of flowering (June). The rose to white petals vary from 12 to 18 in number, each extending ¾ to 1 inch long. Bitterrroot is found in dry, open and often stony soil.

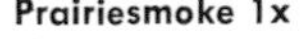

Prairiesmoke 1x

Pink Wintergreen 1x

Holboell Rockcress 1x

Bitterroot Lewisia 1x (Stockert)

JAMES SAXIFRAGE Saxifrage Family

Telesonix jamesii

Beautiful rose-pink flowers in dense clusters, kidney-shaped leaves and growth in rock crevices characterize this subalpine to alpine herbaceous plant. When viewed under a hand magnifier, a number of other features about the plant become apparent. The stem, which is 6-10 inches high is covered with glandular hairs and often becomes purplish. The 10 stamens are topped by black anthers. Finally the seeds are brown and shining. When the mountaineer sees this plant growing in inhospitable rock cracks, he knows the summit is not far.

GLOBEMALLOW; MOUNTAIN HOLLYHOCK Mallow Family

Iliamna rivularis

Along streams and roadsides from Mammoth Hot Springs south to Teton Pass the large, pink flowers add their distinctive color during July and August. The stems are stout, branched and reach a height of 4 or 5 feet. The maple-like leaves are 2 to 8 inches across, generally with 5 lobes. The individual flowers are up to 2 inches broad and resemble the cultivated hollyhocks. Irritating hairs cover the fruits which break open like segments of oranges.

LONGLEAF PHLOX Phlox Family

Phlox longifolia

The leaves of many western species of *Phlox* are needle-like, but this species has the longest leaves. The tube of the corolla is from ½ to ¾ inch long and this is crowned by lobes which extend at right angles. Petal color is variable, but it is usually some shade of pink. Look for this species in sagebrush communities where the soil is sandy and gravelly. The stems often exceed 4 inches in length. Many species of the *Phlox* have been brought into cultivation especially for rock-garden use. Longleaf Phlox occurs from British Columbia south to Arizona and New Mexico.

James Saxifrage 1x

Globemallow 1x

Longleaf Phlox 1x

NORTHERN SWEETVETCH Pea Family

Hedysarum boreale

Since the flowers of this species closely resemble those of locoweeds and milkvetches, it is wise to look for the fruit pods to verify identification. Sweetvetch pods are flattened and constricted between the seeds so that each section appears almost round. The other mentioned plants have pods shaped like garden peas. Inhabiting dry, rocky soil in both parks, this plant is found from 6,700 to to 8,500 feet. The seeds are eaten by small rodents; even the roots are edible and were eaten by the Indians. The keel of the flower is nearly straight and longer than the wings.

RED CLOVER Pea Family

Trifolium pratense

This European clover is widely naturalized and cultivated. In the park it has established itself in wet meadows and along streams in such areas as Old Faithful and the Oxbow Bend of the Snake River. The many species of clover can be recognized as herbaceous plants with palmate leaves divided into 3 leaflets and flowers clustered into dense heads or short spikes. The hand lens reveals that the uppermost petal does not usually stand erect, but is folded lengthwise over the wings and keel. The fruit pods are short and mostly concealed in the calyx. Clovers are high in protein and can be eaten raw, but sparingly as they may be hard to digest.

STEERSHEAD Bleeding-heart Family

Dicentra uniflora

One of the most beautiful harbingers of spring, the Steershead is unique in its flower structure. Like other bleeding-hearts it has 2 sepals, 4 petals, and 6 stamens. Only 2-3 inches high, the plant has a single flower at the tip of each leafless stem. The longer outer petals are curved backward, exposing the tops of the inner petals. The whole flower is about ½ inch long. Growing in gravelly soil, the plant is in the rare category and should be protected. Members of the genus have poisonous alkaloids, but because of their small size losses of animals are never serious.

Northern Sweetvetch 1x

Red Clover 1x

Steerhead 2x (Lane)

SUBALPINE SPIREA

Rose Family

Spiraea densiflora

The Subalpine Spirea is quite distinctive and a much more attractive plant than the white flowered species, *S. betulifolia.* Hiking the canyon trails, the plant enthusiast will find this shrub usually besides the fast moving streams above 8,000 feet. It has a branching habit and reaches a height of 4 feet. The rose-pink flowers are sweet-scented and form flat-topped clusters about 1 or 2 inches across. The flower heads are soft and fluffy in appearance due to many long, interlacing stamens of adjoining blossoms. The leaves are simple and bear small teeth on the margins.

ALPINE LAUREL

Heath Family

Kalmia microphylla

Valley or subalpine lake shores are the perfect setting for this small, straggling, evergreen shrub which is less than 1 foot in height. The opposite leaves are leathery in texture, and the margins are rolled under so that they appear narrowly linear. From the time flower buds appear until complete flower expansion occurs, the various changes are a delight to behold. At one stage the bud resembles a blob of cake frosting. The saucer-shaped flowers have 10 small stamens held in pockets; and when an insect visits the flower, its body is showered with pollen. The visitor then flies to another flower and cross-pollination is assured.

RED WILLOWHERB

Evening Primrose Family

Epilobium latifolium

Like other evening primroses, the individual flowers of this species develop on a plan of four — 4 sepals, 4 magenta petals, 8 stamens, and 4 stigma lobes. The 3 to 12 flowers are in short racemes with leafy bracts. This species inhabits moist subapline meadows and flowers from late July to early September. The seed pods are slender, 4-celled capsules with numerous seeds each bearing a tuft of soft, white hairs at the tip.

Subalpine Spirea ¾ x DO

Alpine Laurel 1x

Red Willowherb 1x

SLIMPOD SHOOTING STAR Primrose Family

Dodecatheon conjugens

The Shooting Star is similar to the cultivated Cyclamen, having flowers in umbels and petals which are reflexed backward. The flowering stems grow to 12 inches tall, and the flowers are from ½ to 1 inch long. Careful examination of this striking flower with a hand lens, will reveal that the stamens are opposite the corolla lobes, a feature quite different from most families. The slimpod Shooting Star is commonly a plant of the sagebrush plain or the moist mountain meadow. The several western species are difficult to separate. Flowering may begin in early June in the valleys.

LADYSTHUMB KNOTWEED Buckwheat Family

Polygonum amphibium

Late in the summer, dense rose-colored spikes of Ladysthumb Knotweed may be seen in lakes, beaver ponds or slow moving streams. These clusters of flowers have reminded some visitors of a painted fingernail. This plant is a perennial with floating or submerged stems. Each flower has from 4 to 6 sepals and from 4 to 9 stamens. The leaves are oblong-elliptic, smooth, and from 2 to 4 inches long. As the scientific name implies, it is at home on land or in the water; however, this form is usually aquatic.

LEWIS MONKEYFLOWER Figwort Family

Mimulus lewisii

Hiking the high canyon trails from 7,000 feet to 9,000 feet, one is very likely to see the Lewis Monkeyflower growing very close to some small stream. It was scientifically named after Lewis of the Lewis and Clark Expedition. The 5 petals of the flower are united into a tube with spreading corolla lobes. Such structure is well adapted to pollination by bees. When a bee crawls into the wide opening of the corolla for nectar, its back becomes dusted with pollen which it carries to the next flower. As the bee crawls into the tube of the second flower, its back brushes pollen onto the stigma.

Slimpod Shooting Star ¾x

Ladysthumb Knotweed ¾x

Lewis Monkeyflower 1x

MOUNTAIN SNOWBERRY Honeysuckle Family

Symphoricarpos oreophilus

The snowberry is an erect shrub with numerous, slender twigs, and it is one of the few woody species with large, white berries. The simple leaves are opposite and oval in outline. The flowers are small and more or less bell-shaped, arranged in rather dense terminal clusters. The flower parts of the individual flowers vary from 4 to 5, but the petals form a corolla tube. Snowberries are an important wildlife food in the western parks. The fruits ripen in late August and frequently remain available on the bushes for many weeks. These two-seeded berries are especially valuable as food for grouse and songbirds. The foliage is occasionally eaten by deer. An extract of the roots was used by the Indians for the treatment of colds and stomach-ache. The odor from the flower is unpleasant.

LYALL'S ROCKCRESS Mustard Family

Arabis lyallii

Members of the Mustard Family are distinctive because of their cross-shaped flowers formed by 4 petals, 4 sepals and 6 stamens. Also called Crucifers, these plants have flowers in racemes and usually 2-chambered fruits. The Rockcress pictured represents a difficult genus, the species of which cannot be identified without fruits and basal leaves. Look for this perennial in the alpine ecosystem on rocky ridges and exposed slopes.

RUSTY-LEAF MENZIESIA Heath Family

Menziesia ferruginea

This erect shrub, reaching up to 8 feet in height, is locally common around the shores of valley lakes in Grand Teton and some steep slopes of Mt. Moran. The thin leaves form rosettes at the ends of slender, erect branches. The plants have a similar appearance to huckleberry plants especially in regard to the flower. However, the fruits are dry capsules and not edible, hence the additional common name, Fools Huckleberry.

SPREADING DOGBANE Dogbane Family

Apocynum androsaemifolium

This is a wide branching semi-shrub growing to a height of 2½ feet. The stem and leaves have a milky juice. It is common in the open Lodgepole Pine forest. The numerous, small, bell-shaped flowers with pink striped petals are borne in clusters. Dogbane was long used by the Indians as a heart medicine. The fruit was boiled while still green, and the resulting brew was taken as a liquid drink.

Mountain Snowberry ¾x

Lyall's Rockcress 1¼x DO

Rusty-leaf Menziesia ¾x DO

Spreading Dogbane 1x (Stockert)

RED MOUNTAINHEATH; MOUNTAINHEATHER Heath Family

Phyllodoce empetriformis

This small, mat-forming, evergreen shrub will often greet the wilderness hiker when he rests at the shore of subalpine lakes. With their crowded, needlelike leaves and urn-shaped flowers, the species of *Phyllodoce* resemble the heathers of Europe. *P. glanduliflora* grows in a similar habitat but can be easily recognized by the yellow-green corollas. Note how the flowers always nod and the stamens are well hidden within the corolla. Late July and August are the months in which to find the blossoms.

WATER SMARTWEED Buckwheat Family

Polygonum coccineum

Like the Ladysthumb Knotweed this plant blooms in late summer and inhabits small ponds that tend to dry up in August. The handsome spikes of pink flowers are from 2 to 3 inches long. Careful observation of a single flower reveals 4 to 6 sepals and 2 to 3 styles which rise above the sepals. The lanceolate leaves form a sheath around the stem. Leaves, roots and seeds of smartweeds may be used as edible food in case of wilderness emergency.

FIREWEED Evening Primrose Family

Epilobium angustifolium

The common name of this plant refers to its ability to populate burned-over and logged areas with a beautiful cover of deep pink flowers. The stem, bearing many alternate, lance-like leaves, may reach a height of 5 feet. The slightly irregular flowers are borne in a long, slender inflorescence which may be a foot in length. In the fall these inflorescences take on a fluffy, white appearance because the seeds are provided with long, white hairs. Flower parts are in fours or multiples of fours.

SHORTSTLYE ONION Lily Family

Allium brevistylum

Wild onions have been used for their edible bulbs since ancient times, both in the New and Old Worlds. The Indians ate the bulbs raw or cooked them with other food. Many mammals, such as bear and elk, also utilize these odoriferous plants. There are about 500 species of onions in the world, and all have the same distinctive flower structure — 3 sepals, 3 petals, 6 stamens and 3 fused carpels. The flowering stem is commonly 6 to 13 inches, and the flower cluster has several tissue-like bracts where the pedicels join the main stem. Flowering begins in June and continues into August at higher elevations.

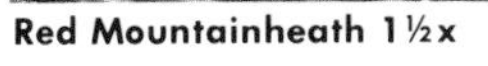

Red Mountainheath 1½x DO

Water Smartweed 1x

Fireweed ½x DO

Shortstyle Onion 1x DO

PRINCESPINE PIPSISSEWA Heath Family

Chimaphila umbellata

This plant is a trailing and somewhat woody perennial with leafy shoots and flowering branches. In our parks we find it in the shade of the woods which surround the many lakes. The narrowly wedge-shaped leaves are evergreen and have margins with forward pointing teeth. The flowers, borne in small, terminal clusters, are pink and contain 10 stamens and a conspicuous stout style. The fruits are roundish capsules holding numerous small seeds. The name, pipsissewa, is evidently of Indian origin, and the plant was used for a wide variety of ailments, such as rheumatism and fevers. The flowers appear in late July or early August.

NARROWLEAF COLLOMIA Phlox Family

Collomia linearis

Disturb a coniferous forest with a bulldozer and one of the first annual species to invade will be the Narrowleaf Collomia. The stem varies from 4 to 15 inches high according to moisture. The pink, tubular flowers may reach ½ inch long and form dense clusters in axils of leafy bracts. The calyx tube is papery in texture and enlarges as the fruit matures. The generic name *Collomia* means glue and refers to the mucilaginous quality of the moistened seeds.

MOSS CAMPION Pink Family

Silene acaulis

The small branches of this perennial plant form a tightly interwoven cushion connecting to a deep penetrating taproot. In the alpine ecosystem such a growth habit raises the temperature inside the cushion and helps to create a microclimate which is more suitable for survival. Sometimes in the older cushions, seeds of other plants germinate and gradually become established. At first glance the plant may seem to resemble a dwarf *Phlox,* but the petals of the Moss Campion are distinct and separate, whereas they are united in the *Phlox.* Also known as Moss Pink, this plant may be 10 years old before it begins to flower. In both parks this species inhabits rocky sites above 9,800 feet.

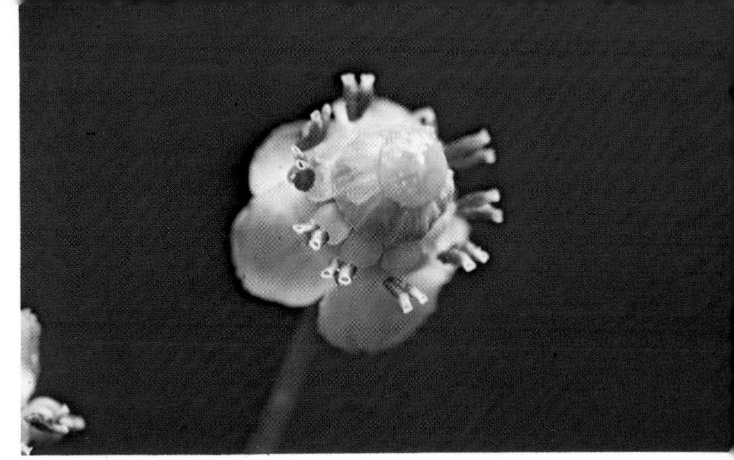

Princespine Pipsissewa 2x

Narrowleaf Collomia 1x

Moss Campion 1¼x Mortensen

FAIRYSLIPPER; CALYPSO ORCHID Orchid Family

Calypso bulbosa

Yellowstone and Grand Teton have at least 15 species of orchids, but this is the most beautiful and striking. It is found in cool, deep shaded areas during the first three weeks of June. Usually it has only one small, green, basal leaf, which, along with the stem, arises from a bulb embedded frequently in decaying wood or organic matter. The flower resembles a small lady's slipper with its cup-like lip. This plant needs protection throughout the country because of its rapidly disappearing habitat requirement and its delicate beauty which causes people to pick first and think afterwards.

ROSE PUSSY TOES Composite Family

Antennaria microphylla

All of the plants which are called pussy toes are recognized by white, woolly hairs on the stems and leaves and several rows of papery white or colored bracts. These remain on the flowering heads indefinitely and are responsible for another common name, Everlasting. Many *Antennaria* species form seed without the process of pollination. This results in the formation of numerous races and leads to confusion in identification. Rose Pussy Toes can be found in dry, open places or open woods up to about 8,500 feet. The bracts vary from white to rose pink.

PARRY LOUSEWORT Figwort Family

Pedicularis parryi var. *purpurea*

While hiking above 10,000 feet, wilderness enthusiasts will be thrilled to see this low plant (2-3 inches) of the tundra. The corolla is markedly two-lipped. Two petals make up the upper lip, and together they are strongly arched and compressed to protect the stamens. Three petals are fused to form the lower lip. The basal leaves are pinnately divided or cleft and closely resemble a fern leaf. Individual flowers are ½ inch long and appear in August and early September. Bees are responsible for pollination.

Fairyslipper 1x

Rose Pussy Toes 1x

Parry Lousewort 1x

ELEPHANTHEAD Figwort Family

Pedicularis groenlandica

Close examination of a single flower will reveal why this plant has such a descriptive common name. The upper lip of the corolla has a long, upcurving beak. Two petals of the lower lip are shaped like ears. Together the parts of this irregular flower have an amazing resemblance to an elephant's head. The leaves are all pinnately divided. Look for the purple spikes in wet meadows from 6,700 feet up to 9,000 feet during July and August.

COMMON CATTAIL Cattail Family

Typha latifolia

The Common Cattail is probably the most famous of all the edible wild plants. Rootstocks, young shoots, and young flower stalks can be utilized as delicious wild food. The tall, brown "cat's tail" is composed of thousands of minute flowers, each one being little more than a single pistil. Above the pistillate spike temporary staminate flowers appear. When the pollen is shed, these upper flowers disappear. Rootstocks are high in starch content and this carbohydrate can be extracted to yield a flour comparable to that made from wheat or corn.

LEOPARD LILY; PURPLESPOT FRITILLARIA Lily Family

Fritillaria atropurpurea

Leopard Lilies are easily passed by if the park visitor isn't watching for them. The fast growing stem (up to 20 inches tall) bears 2 to 4 nodding flowers. When one of these is examined closely, the yellowish spots are revealed against a bronze or dull purple background. The odor from such flowers is very unpleasant to humans, but very attractive to flies which are probably responsible for pollination. The plant grows in open forests or grassy slopes; it flowers in June.

Elephanthead 1x

Common Cattail ⅓x

Leopard Lily ¾x

COMMON INDIAN PAINTBRUSH Figwort Family

Castilleja miniata

The actual flowers of the plant are narrow, tubular, and greenish-yellow. The vivid scarlet of the leafy bracts provide the color of the most common species, and yet there are several species in the parks whose bracts are white, yellow, orange, or pink. The plants bloom from mid-June to early September and are found from the Snake River bottom to 11,000 feet. Wyoming has chosen one species as its state flower (*C. linariaefolia*).

RHEXIA-LEAVED PAINTBRUSH Figwort Family

Castilleja rhexifolia

Identification of the park's 10 to 12 species of *Castilleja* can be quite frustrating, and it really takes a botanical specialist working with technical details to separate them. Minute details about the bracts, calyx and corolla must be considered. In all *Castilleja* species the corolla has a narrow, folded upper lip, called the galea, and a lower lip with 3 lobes or teeth. *Castilleja rhexifolia* reaches up to 12 inches and bears leaves which are narrowly lanceolate. It inhabits high mountain meadows and blooms during July and August.

BROWNS PEONY Peony Family

Paeonia brownii

This plant is infrequent in both parks; therefore, there should be some excitement in discovering it. Sagebrush or open coniferous forests are the habitats to search. Browns Peony is somewhat succulent and has several stems each bearing several dissected leaves. The single flowers are borne at the tips of curved stems. There are usually 5 or 6 concave sepals blending with 5 or 6 reddish petals with yellowish margins. The stamens are numerous and fit tightly around the 4 or 5 pistils which become follicles. Blooming time is mid-June.

Common Indian Paintbrush 1x

Rhexia-leaved Paintbrush ¾ x

Browns Peony ¾ x

WESTERN CONEFLOWER Composite Family

Rudbeckia occidentalis

August is the month to watch for this stout herb which varies between 2 to 5 feet high. The flowering heads are on tall, leafy stems. The colorful ray flowers seen in other composites are lacking in this species, and yet the numerous tiny disk flowers are obviously displayed on a cylindrical cone nearly 2 inches long. The plant can be found in moist, shady places such as Gibbon Meadows and Tower Falls. In Grand Teton it grows along the highway north of Colter Bay.

MARSH CINQUEFOIL Rose Family

Potentilla palustris

The common name cinquefoil means "five leaves" and is used for all species regardless of the number of leaf segments. The scientific name *Potentilla* means "the little potent one" because of the supposedly medicinal value of one species. *P. palustris* is the only species with red petals. The 20-40 inch stems may be prostrate or floating in water. *Palustris* means "marshy or boggy" and this species is found at the edge of lakes (north shore of Shoshone Lake in Y.N.P. and Bradley Lake in G.T.). The roots are edible either boiled or roasted.

MOUNTAIN TOWNSENDIA Sunflower Family

Townsendia alpigena

The Townsendias are perennials with deep tap roots. The leaves are mostly narrow or spoon-shaped and the alpine species illustrated has a dense coating of hairs to reduce water loss and possibly to protect from ultra-violet rays. The flower heads obviously resemble those of Asters and Daisies, but a specialist separates this genus on the basis that the Townsendias have flattened bristles atop the fruit, whereas those of Asters and Daisies are round and hairlike. Look for this attractive composite on slopes and ridges in the alpine ecosystem especially on the summit of Rendezvous Mountain during August.

Western Coneflower 1x

Marsh Cinquefoil 1x

Mountain Townsendia 1x

STRIPED CORALROOT Orchid Family

Corallorhiza striata

The Coralroot Orchids are devoid of the green pigment, chlorophyll, and cannot manufacture their own food. All their nourishment comes from decaying organic matter in the soil and is absorbed by coral-like underground stems. The flowers are numerous on an 8 to 15 inch leafless stem. In this particular species the broad lower petal is almost completely purple, but the upper petals and sepals have 3 dark purple stripes.

WOODLAND PINEDROPS Heath Family

Pterospora andromedea

Since Pinedrops lacks well-developed leaves and the pigment, chlorophyll, it cannot produce its own food; and, therefore, takes its nourishment from other sources. Recent research has shown that this plant lives as a parasite on soil fungi. The reddish-brown stem may reach a height of 3 feet, and it is covered with sticky hairs. The bell-shaped, nodding flowers bloom from the bottom upward, commencing in July and lasting into the middle of August.

SKYROCKET GILIA Phlox Family

Gilia aggregata

This plant adds the most color to the sagebrush flats during the month of July, but it can also be found in open wooded areas. Usually this species is a biennial, producing only a small clump of basal leaves the first year, followed by a 1½ to 2½ foot flowering stock the second year. The flaring corolla lobes are bright red with yellowish mottling on the inside. The flowers are especially attractive to humming birds which thrust their bills down the tube of the corolla seeking nectar at the base. In the hovering and collecting process the bird's head becomes covered with pollen and when it hovers at the next flower, pollination is assured.

MOUNTAIN LOVER; MYRTLE PACHISTIMA Staff-tree Family

Pachistima myrsinites

The only representative of this family is common in Grand Teton and may eventually be found in Yellowstone west of the Continental Divide. This evergreen shrub has the appearance of low growing boxwood and grows in the shade of Aspen or Lodgepole Pine. The opposite leaves have serrate leaf margins. The diminutive flowers are red, saucer-shaped and constructed on a plan of four. One can remember the scientific name of this species by repeating the simple saying, "Pa kissed ma for mercenary reasons."

Striped Coralroot 1x  DO

Woodland Pinedrops ⅔x 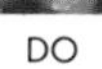 DO

Skyrocket Gilia 1x

Mountain Lover 2x

WESTERN CORALROOT — Orchid Family

Corallorhiza mertensiana

The 5 species of the genus *Corallorhiza* that are known to be in the parks are a stimulating challenge to discover. While the generic name means "coral root" and refers to the underground portion of the plant because it resembles coral, it is not a root at all, but rather a brittle, branched rhizome. The flowering stem which grows from the rhizome is lacking in chlorophyll and is surrounded by several scale-like vestigal leaves. The flower of this species has a lip with small teeth on each side in the basal half. Like other coralroots this species is found in decaying plant material in the coniferous forests and blooms in late June and early July.

SPOTTED CORALROOT — Orchid Family

Corallorhiza maculata

Like the Striped Coralroot these orchids cannot manufacture their own food, and all nourishment comes from decaying organic matter in the soil of the Lodgepole Pine stands. Slender asparagus-like spears appear in late June or early July and quickly develop racemes from which the flowers open. The specific name, *maculata,* means "spotted" and is given because of the purple to brown spots on the lip of the flower. Albino and partially albino individuals are occasionally encountered. If one uses a hand lens, a small projecting tooth can be seen at the base of the lip.

WILD BLUE FLAX; LEWIS FLAX — Flax Family

Linum perenne var. *lewisii*

Wild Flax is found in dry, rocky soil, but never in any great quantity. The rather numerous flowers are located on very slender stems which bend and bow to every passing breeze. The 5 blue petals are extremely fragile and will fall off at the slightest handling. Unthinking flower pickers will be sadly disappointed five minutes after plucking a handful of these delicate plants. The blossoms open early in the forenoon and usually close in the late afternoon. Seeds are formed in small, hard capsules which split open late in the fall.

Western Coralroot 3x

Spotted Coralroot 1x

Wild Blue Flax 1x

HAIRY CLEMATIS; SUGARBOWL Buttercup Family

Clematis hirsutissima

This striking, herbaceous plant may grow with the Big Sagebrush where its deep purple sepals will be in sharp contrast to its dull environment. The flower has no petals but has many stamens and pistils. The style of the pistil lengthens into a plume as the fruits mature. Because of the thickness and hairs on the sepals, this plant has been called Leather Flower. The species name *hirsutissima* means very hairy. Another feature helpful in identification is the leaf which is 2 to 4 times pinnately dissected. The plant blooms in late June.

ROCK CLEMATIS Buttercup Family

Clematis occidentalis

This woody-stemmed clematis species is one of the few climbing vines of the two parks; it has many features to set it apart from the other woody plants. The leaf stalks or petioles act like tendrils and attach themselves to some support by twining. The flowers lack true petals, but are conspicuous because of the purple, petal-like sepals and numerous stamens. The fruits which develop from numerous pistils have long, fuzzy styles. Sometimes these fruit clusters are 2 inches in diameter. The Indians and early settlers used an infusion of stems and leaves for the treatment of colds and cuts. The generic name, *Clematis,* is the ancient name given by Dioscorides to a climbing vine. Clematis Gulch at Mammoth Hot Springs is named for this species. Another climbing vine, *Clematis ligusticifolia,* has smaller white flowers and is found in Death Canyon of Grand Teton.

SKY PILOT Phlox Family

Polemonium viscosum

Sky Pilot is truly a symbol of lofty alpine habitats. This and other *Polemonium* species have one striking feature in common — a strong, skunky odor. This strong fragrance is traceable to the sticky, glandular hairs that cover leaves and stems. The blue-violet corolla is funnel-shaped and contrasts well with golden-orange stamens. Flowering occurs in July and August creating patches of blue among the rock of alpine moraines, such as the Middle Teton Glacier Moraine. The leaves are up to 6 inches long and pinnately compound; each leaflet is divided into 3-5 lobes. A valley dwelling species, *P. occidentale,* grows in wet meadows of the National Elk Refuge.

Hairy Clematis 1x

Rock Clematis 1x

Sky Pilot 1½x

DO

SPOTTED KNAPWEED Composite Family

Centaurea maculosa

This introduction from Europe is now wide spread throughout North America and seems to be making an invasion in both parks especially along roadsides. Where it does establish itself, there will be a surprising display of pink to purple color during August. This biennial to perennial plant is conspicuously branched and may reach a height of 3 feet. The leaves are pinnately compound. This species may be confused with some *Aster* species, but it is easily distinguished by the outer involucral bracts which have blackish tips and finely divided margins. The specific name, *maculosa,* means spotted, referring to sticky spots on the leaves.

CANADA THISTLE Composite Family

Cirsium arvense

This species is a cosmopolitan, noxious weed from Eurasia and, hence, the common name is misleading. The plant is comparatively slender with numerous small heads. The pinkish-purple flowers add a decorative touch to some roadside cuts. Thistles were often used as food by the Indians; and in spite of their spiny exterior, can provide food, especially in times of emergency. Young leaves, tender roots, or flower heads may be used. The two parks have about 8 species, some of which are difficult to identify.

SHOWY FLEABANE Composite Family

Erigeron speciosus

Separating species of *Erigeron* from *Aster* can be frustrating, but generally many fleabanes flower earlier in the summer season than Asters. Also the ray flowers of *Erigeron* are more numerous and narrower. Showy Fleabane stems may be up to 25 inches and may bear nearly a dozen flower heads, each with from 70 to 140 slender whitish to bluish ray flowers. The tubular disk flowers are yellow-orange. The plant can be found in dry to moist soil in open, wooded areas.

Spotted Knapweed ¾x

Canada Thistle 1x

Showy Fleabane 1x

PACIFIC ASTER Composite Family

Aster chilensis

In the genus *Aster* we find many variable species which intergrade with one another; even botanical specialists would be cautious about stating the number of species in the two parks. A conservative estimate would be between 12 and 18. The species illustrated here has a downy stem. The lowest leaves have basal lobes which extend around the stem. The numerous flowering heads have green tipped bracts and up to 35 ray flowers ½ inch long. The ray flower color varies from violet to pink or almost white. Flowering extends from late July through mid-September, and the plant occupies dry, open places.

THICKSTEM ASTER Composite Family

Aster integrifolius

While many species of asters are hard to separate, the Thickstem Aster has some distinctive features that set it apart. The rather tall stem (10-24 inches) has only a few flower heads in a narrow inflorescence. The sparse, purple ray flowers vary from 8 to 20 and surround the yellow-orange disk flowers. The involucral bracts below the flowers are covered with glandular hairs. Late August and early September are less drab because of this ragged, beautiful perennial.

ALPINE ASTER Composite Family

Aster alpigenus var. *haydenii*

This dwarf perennial is always found high in the mountains clinging to life in cracks of weathering rocks. In the flowering head there are from 10 to 40 ray flowers about ½ inch long. The involucral bracts, as in all asters, form several overlapping rows of different lengths and may be purplish. The basal inner leaves are somewhat folded and are up to 8 inches long. There are 3 varieties in this species, and our variety, *haydenii,* is the smallest of the three.

Pacific Aster 1x (Stockert)

Thickstem Aster 1x

Alpine Aster 1x

EVERTS THISTLE Composite Family

Cirsium scariosum

Large numbers of brown, spotted beetles are attracted to the numerous, composite, purple heads. The thick, fleshy stems are covered with long, arching leaves whose margins are covered with weak spines. Along the Jackson Hole Highway, the visitor is sure to see many plants of this species rising as much as two feet above the grasses of the meadows. Truman Everts, one of the early explorers of Yellowstone, was lost for over a month in 1870 and is said to have subsisted mainly on the root of this thistle.

COLUMBIA MONKSHOOD Buttercup Family

Aconitum columbianum

The upper sepal of this irregular flower is modified into a hood-shaped structure and is responsible for the common name. In addition there are 2 broad sepals at the side and 2 small sepals below. Concealed within the hood are 2 petals and numerous stamens. The stem is stout and varies in height from 2 to 5 feet. Monkshood inhabits wet meadows and stream banks up to 9,000 feet in the major canyons. The flowers are generally purple but occasional albinos occur. Late June through August.

LOW LARKSPUR; UPLAND LARKSPUR Buttercup Family

Delphinium nuttallianum

Although it has practically no effect on park wildlife, the Upland Larkspur is one of the most serious of the poisonous plants of the western cattle ranges. It is similar to Duncecap Larkspur except that it grows on foothills and sagebrush flats rather than in wet places under the aspen. Also, it is only about a foot high compared to 3 to 6 feet for Duncecap Larkspur. The flowers have 5 irregularly shaped, dark blue sepals enclosing 4 much smaller petals of lighter color.

DUNCECAP LARKSPUR Buttercup Family

Delphinium occidentale

The Duncecap Larkspur blooms during late July and early August in wet places, whereas, the Low Larkspur flowers on the sagebrush flats during June. Both species are easy to recognize as larkspurs, because there are 5 petal-like sepals, the upper one spurred, and 4 petals. The stems are stout, upright, and often reach a height of 6 feet. The leaves are compound, with the segments radiating from the basal end like the fingers of a hand.

Everts Thistle $\frac{1}{3}$x

Columbia Monkshood $\frac{3}{4}$x

Upland Larkspur $\frac{3}{4}$x

Duncecap Larkspur $\frac{3}{4}$x

MOUNTAIN BLUEBELL Borage Family

Mertensia ciliata

Reaching a height of 3 feet or more, this showy Mountain Bluebell inhabits subalpine areas particularly along streams. The alternate leaves and tender stems are enjoyed by many animals of the parks. The tubular flowers are purplish in bud, but rapidly turn blue as the blossom opens to full size. Five stamens are attached to the inside of the corolla. Look for this species on the trail to Lake Solitude or while traveling the highway over Dunraven Pass. The peak blooming time is late July and early August.

THISTLE MILKVETCH Pea Family

Astragalus kentrophyta

Thistle Milkvetch occurs from sandy deserts and badlands to alpine ridges and talus; our variety, *implexus,* has a purplish banner (upper petal) and inhabits alpine environments. The low plants form mats among the rocks, and each hairy leaflet has a sharp spine at the tip. This is indeed a confusing genus having about 1,500 species and some of them are extremely poisonous to livestock. Considering the abundance of species, it is not surprising that the Indians found some species were edible, particularly seed pods and roots. However, the inexperienced plant gatherer should be careful to avoid the many poisonous species.

MOUNTAIN PENSTEMON Figwort Family

Penstemon montanus

One common name for this group, Beardstongue, refers to the presence of a sterile stamen ("tongue") in addition to 4 fertile stamens. No pollen is formed by the sterile stamen, but it is commonly covered with a "beard" or tuft of hairs. There are over 200 species of *Penstemon* in western North America, but the identification is difficult and requires careful observations of minute flower details. The Penstemons, in general, are found in open, rather dry and rocky habitats. The species pictured is most often seen in the alpine ecosystem, but may be found as low as 8,000 feet in Yellowstone. The anthers are woolly with tangled hairs.

Mountain Bluebell ¾x

Thistle Milkvetch 2x (Stockert)

Mountain Penstemon 1x

COMMON BLUE-EYED-GRASS Iris Family

Sisyrinchium idahoensis

Whether seen from the board walks at Old Faithful or along the banks of the Snake River, these miniatures of the Iris Family will always draw favorable comments. The flattened stems are about 6 to 12 inches tall and are topped with 1 to 5 flowers. Where the 3 sepals and 3 petals join there is usually a yellow center. Note also that each perianth member is tipped with a minute point. These plants grow in moist soil and bloom in July or August.

FRINGED GENTIAN Iris Family

Gentiana detonsa

The Park Service chose this plant as the official flower of Yellowstone National Park. In this area it is common and blooms throughout the tourist season, beginning in June in the warm earth of the geyser basins. However, in Grand Teton National Park it occurs in rather limited areas. Look for flowers along the main highway between Jackson Lake Dam and the Jackson Lake Lodge. The petals are fused into a corolla about 2 inches long; the lobes of which are fringed.

ROUNDLEAF HAREBELL Bluebell Family

Campanula rotundifolia

Through July and August, this delicate, herbaceous perennial is abundant in the coniferous forests and along roadside cuts. The specific name *rotundifolia* refers to the roundish, heart-shaped, basal leaves. While these wither early, the narrow, pointed stem leaves remain. A conspicuous feature of the flowers is that, although the buds grow erect, the open blossoms droop or are horizontal, giving protection to the pollen from the rain. Occasionally, completely white or albino flowers will grace the stems of this widespread plant. In rock gardens this plant spreads rapidly by underground stems.

FIELD MINT Mint Family

Mentha arvensis

The mint family can be immediately recognized by three characteristics — irregular flowers, square stems and opposite leaves. But because the flowers of most species are small, it is hard to separate the genera without checking minute details of stamens and other technical features. The flower clusters of Field Mint are in the axils of foliage leaves. The corolla is four-lobed, the upper, notched lobe is usually broader than the others. The plants of this genus secrete aromatic volatile oils used as medicines and flavoring agents, i.e. menthol, peppermint and spearmint. Wet woods and streambanks are typical habitats for this species.

Common Blue-eyed-grass 1x

Fringed Gentian 1x

Roundleaf Harebell 1x (Kretzer)

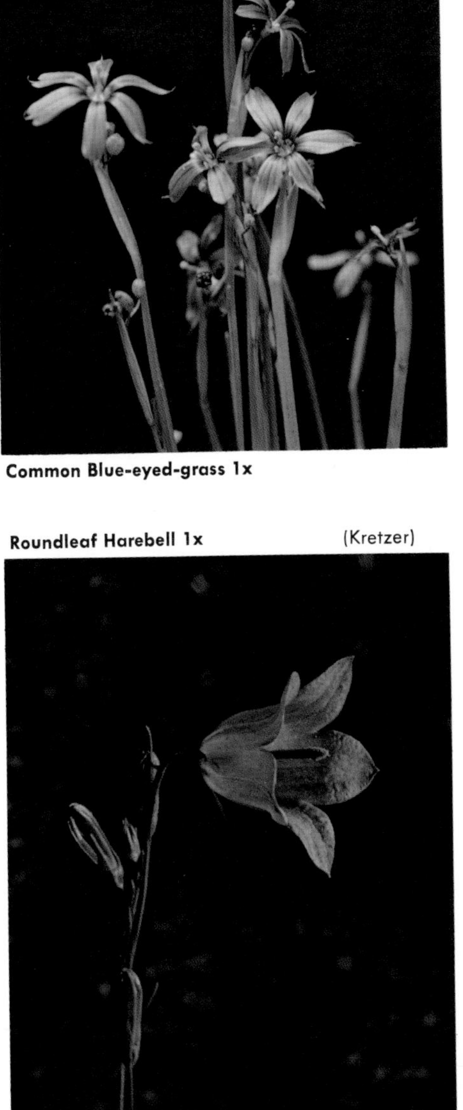

Field Mint 1x (Stockert)

SILVERY LUPINE Pea Family

Lupinus argenteus

Nearly 600 species names have been proposed for this genus in North America. The difficulty arises from the fact that many species hybridize yielding a broad spectrum of intergrading forms. The plants usually grow in dense, colorful clumps in the sagebrush as well as under stands of Lodgepole Pine. Flower photographers will find they provide fine subjects throughout most of June, July and August. The leaflets of each leaf radiate out in a finger-like fashion.

COMMON SELFHEAL Mint Family

Prunella vulgaris

Common Selfheal was once esteemed for healing wounds, but now it is considered valuable only as a refreshing beverage which can be made by chopping and boiling the leaves. The corolla of the flower is two-lipped, the upper lip forms a hood. In between the flowers of the spike one can see green bracts. The flower color is variable from bluish to violet. There are 4 stamens in two sets. The plant inhabits stream banks, lake shores or moist meadows and flowers in July or August.

MOUNTAIN BOG GENTIAN Gentian Family

Gentiana calycosa

As the common name implies, this subalpine plant is limited between 7,000 to 10,000 feet in the mountain canyons, especially along stream banks. The stems are about 5 to 13 inches high and possesses opposite leaves with smooth margins. Usually there is a single flower about 1½ inches long, the interior of which is spotted with grayish-green dots. Note also that between the major flower lobes are short, double pronged projections. Blooming begins in late July and continues into September in the higher elevations.

Silvery Lupine 1x

Common Selfheal ¾x

Mountain Bog Gentian 1x

MANY-FLOWERED STICKSEED Borage Family

Hackelia floribunda

Also known as Forget-me-not, the Many-flowered Stickseed receives its name because of the hooks on the nutlets that adhere to the fur and clothing when the plant goes to seed. This short-lived biennial grows from 2-3 feet tall and bears numerous bright blue flowers on curving flower stalks. The corolla has a short tubular section which abruptly spreads into 5 lobes. Inside and below the yellow center are 5 small stamens attached to the tube of the corolla. This species is most apt to be found in moist meadows, streambanks and avalanche paths.

SILKY PHACELIA Waterleaf Family

Phacelia sericea

The genus *Phacelia* is large and perplexing with at least 150 species in North America. Silky Phacelia is one of the easiest species to recognize. The stems which reach up to 2 feet have leaves which are pinnately cleft. The flowers form a cylindric inflorenscence. The striking purple stamens extend beyond the ¼ inch corolla tube like long hairs. This perennial plant is found in rather dry soils of trails and roadsides from 6,500 to 8,500 feet and flowers in July and early August. The specific name, *sericea,* means silky and refers to the silvery pubescence covering stems and leaves.

ALPINE SPEEDWELL Figwort Family

Veronica wormskjoldii

Members of the genus *Veronica* have lower, opposite leaves and distinctive flower features such as only 2 stamens and usually 4 petals of unequal size. The species pictured is a perennial mountain plant with pubescent stems and leaves. The common name, Speedwell, refers to the use of some species to cure scurvy. This alpine beauty can be found on Dunraven Pass and in several canyons in the Tetons.

BLUE CAMAS Lily Family

Camassia quamash

This member of the Lily Family is an onion-like plant arising from a bulb which has been used by many Indian tribes as an important food. Wet meadows east of Jackson Lake and meadows adjacent to Yellowstone Lake are the best places to look for this June blooming species. A leafless flowering stalk reaches a height of about 18 inches and is crowned with a loose cluster of purplish-blue flowers, 1 to 1½ inches in diameter. Harrington (see selected references) says the bulbs seem to be lacking in starch, although the sugar content is high. Harrington also reports that many local Indian wars were fought over the collecting rights to certain Blue Camas meadows.

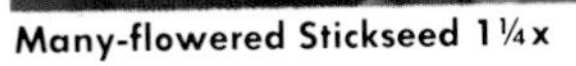

Many-flowered Stickseed 1¼x DO

Silky Phacelia 1x DO

Alpine Speedwell 1¼x DO

Blue Camas 1¼x DO

PARTS OF A FLOWER

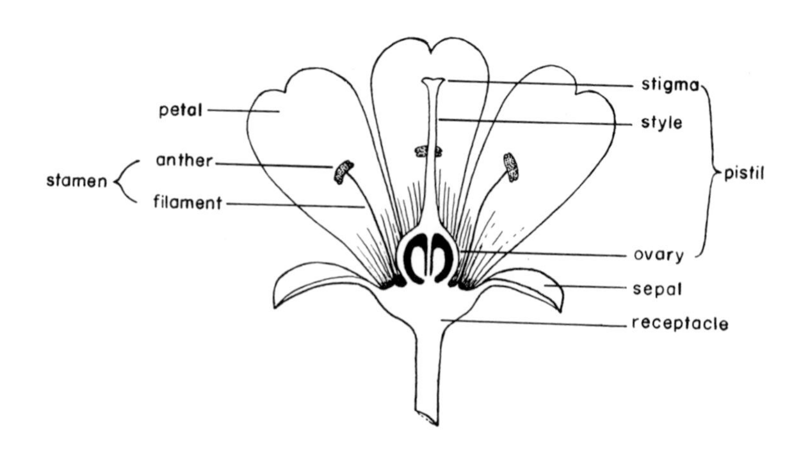

GLOSSARY

ACHENE — A small, dry fruit which does not open by itself.

ANTHER — The pollen bearing part of the stamen.

BIENNIAL — A plant which completes its life cycle and dies in 2 years.

BRACT — A modified leaf associated with a flower.

CALYX — All of the sepals of a flower considered collectively.

CAPSULE — A dry, dehiscent (splitting open) fruit composed of more than one carpel.

CARPEL — The basic unit of a pistil; the pod of a pea is a good example.

CIRCUMBOREAL — Occuring all the way around the northern latitudes.

COROLLA — A collective term referring to the petals of a flower.

DECIDUOUS — Falling off after completion of the normal function.

DISK FLOWER — A central flower of a composite inflorescence (such as the center of a sunflower).

FLORET — An individual small flower of a definite cluster.

FOLIATE — Referring to leaflets of a compound leaf.

FOLLICLE — A dry fruit formed from a single carpel, splitting open along one edge only.

INFLORESCENCE — A flower bearing branch or system of branches.

INVOLUCRAL — Referring to a set of bracts beneath an inflorescence.

NODE — A point on a stem where a leaf is (or has been) attached.

PAPPUS — A modified calyx, usually composed of bristles or awns and always associated with Composite Family.

PEDICEL — The stalk of a single flower.

PEDUNCLE — The common stalk of a flower cluster.

PERIANTH — The collective term applied to the sepals and petals of a flower.

PERENNIAL — Living year after year.

PETALOID — Petal-like.

PETIOLE — A leaf stalk.

PINNATE — Having 2 rows of parts or appendages along an axis, like barbs on a feather.

PISTILLATE FLOWER — A flower bearing 1-more pistils, but no stamens.

POLLINATION — Transfer of the pollen from the anther to the stigma by such agents as wind, insects and birds.

PUBESCENCE — The various types of hairs that cover the surface of a plant.

RACEME — An elongated inflorescence with a single main axis along which stalked flowers are arranged.

RAY FLOWERS — The straped-shaped marginal flowers of the Composite Family; each ray flower is complete with corolla and essential organs.

RECEPTACLE — The tip of a floral axis, bearing the parts of a flower.

SALVERFORM — Having a slender tube and an abruptly spreading set of corolla lobes.

SILIQUE — An elongated capsule of the Mustard Family.

STAMINATE FLOWER — Having 1-more stamens, but no pistils.

SUCCULENT — Fleshy and juicy.

UMBEL — An inflorescence in which the pedicels radiate from a single point like the spokes of an umbrella.

VILLOUS — Covered with long soft hairs.

SELECTED REFERENCES

Baker, Richard G. 1970. *Pollen Sequence from Late Quaternary Sediments in Yellowstone Park*. Science 168:1449-1450.

Craighead, John J., Frank C. Craighead, and Ray J. Davis. 1963. *A Field Guide to Rocky Mountain Wildflowers*. Houghton-Mifflin Co. Boston.

Dorf, Erling 1964. *The Petrified Forests of Yellowstone National Park*. Supt. of Documents, U. S. Government Printing Office, 0-735-958, Washington, D. C.

Harrington, H. D. 1967. *Edible Native Plants of the Rocky Mountains*. University of New Mexico Press, Albuquerque.

Hitchcock, C. Leo, and Arthur Cronquist. 1973. *Flora of the Pacific Northwest*. University of Washington Press, Seattle.

Kirk, Donald R. 1970. *Wild Edible Plants of the Western United States*. Naturegraph, Healdsburg, California.

Long, John C. 1965. *Native Orchids of Colorado*. Denver Museum of Natural History, Pictorial No. 16.

Rickett, Harold W. 1973. *Wildflowers of the United States*. Vol. 6. McGraw-Hill Co., New York.

Weber, William A. 1972. *Rocky Mountain Flora*. Colorado Associated University Press, Boulder, Colorado.

Weiner, Michael A. 1972. *Earth Medicine — Earth Foods*. Collier Books, New York.

INDEX